THE M.I.T. PRESS PAPERBACK SERIES

Candidates, Issues, and Strategies

A Computer Simulation of the
1960 and 1964 Presidential Elections

Ithiel de Sola Pool
Robert P. Abelson
Samuel L. Popkin

THE M.I.T. PRESS
Massachusetts Institute of Technology
Cambridge, Massachusetts

239524

10/25/79 AP

Revised edition, August, 1965

First M.I.T. Press Paperback Edition, August 1965
Third Paperback Printing, August 1970

ISBN 0 262 66003 2 (paperback)
Library of Congress Catalog Card Number: 64-23116
Manufactured in the United States of America

To the memory of the man for whom we worked
J.F.K.

Acknowledgments

In an election every vote counts. In a research project every job counts. We could not have completed a large-scale action-research project such as the one here described without the contributions of many individuals to the endeavor. There were those in the Democratic Party who were farsighted enough to see that advanced scientific research might have a contribution to make to the practical arts of politics. Among them special mention must be made of Thomas Finletter, Robert Benjamin, Robert Kennedy, Michael Nisselsen, Charles Spaulding, Donald Rivkin, Charles Tyroler, Paul Butler, Seymour Peyser, Joseph Baird, Arnold Maremont, and Curtis Roosevelt. Collective mention should also be made of the members of the Executive Committee of the Democratic Advisory Council and of the *ad hoc* groups they assembled to review our work.

The organization whose election project we herein describe is the Simulmatics Corporation. The study is best known as "the Simulmatics Project." To all our colleagues in Simulmatics we extend our thanks for their willing cooperation and support. To the company we are grateful for making our research possible. But thanks to a company is, of course, a euphemism for thanks to some man: in this case the President of the Company, Edward L. Greenfield. Throughout the effort he was every bit as involved as the authors. He managed the project and guided it in all of its practical and political aspects.

Numerous helpers worked on the project long and pressured hours in the days of the campaign while others have helped in reanalyzing the results in the period since. Among them special words of thanks are due George Fishman, George E. McCord, III, Gerald Kramer, Howard Rosenthal, John Forrest, Jr., Helmut Gutenberg, Hayward Alker, James Purcell, Thomas Morgan, and Lewis Dexter. We are grateful to William McPhee for much good advice and to Sidney Furst for conducting a survey for us. The Eagleton Foundation generously provided a grant for postelection reanalysis. It has led to this book. We are grateful to them for making possible the continuation of this work after politically motivated support was no longer in order. The M.I.T. Computation Center was used for this post-election reanalysis. We appreciate their extensive cooperation and the computer time made available.

We have an enormous debt to George Gallup and Elmo Roper, the producers of the main body of public opinion poll data about political behavior in the United States, who have had the statesmanship to make their data available for continuing study through the Roper Public Opinion Research Center at Williamstown,

Massachusetts. Old poll cards are deposited there for research use.

Miss Elly Terlingen and Miss Karen Johnson helped us to turn this manuscript into a published book.

ITHIEL DE SOLA POOL
ROBERT P. ABELSON
SAMUEL L. POPKIN

Cambridge, Massachusetts
April 1964

Contents

1

THE STORY OF THE PROJECT

2

TESTING THE ASSUMPTIONS

3

THE DYNAMICS OF 1960

CONTENTS

4
DO PREDISPOSITIONAL FACTORS SUMMATE?
127

5
POLITICAL TRENDS OF THE 1950'S

6
A POSTSCRIPT ON THE 1964 ELECTION

GLOSSARY
183

INDEX
191

Candidates, Issues, and Strategies

1

The Story of the Project

THE IDEAS BEHIND THE PROJECT

A new social research technique, computer simulation, was put to its first political use during the Presidential campaign of 1960. The simulation, done for the Democratic Party, involved a novel technique for processing public opinion poll data. But the simulation, besides being a step forward in the automating of opinion research, was also a field test of some theories of opinion formation.

Because of its novelty the Simulmatics project has been the subject of a number of sensationalized newspaper and magazine articles and even of a work of fiction.[1] The present report endeavors to correct these lurid fantasies. There was no "people machine"; nor were there superhuman manipulators pulling magic out of computers. Responsible people, not computers, ran the campaign. What was novel was the use of a research technique allowing more intelligent understanding of voter behavior.

[1] Eugene Burdick, *The 480* (New York, McGraw-Hill Book Co., 1964).

Pre-election polling with scientific sampling was first applied to a Presidential campaign in 1936. By 1960, scientific polling, which George Gallup had pioneered a quarter of a century earlier, had become part of the normal arsenal of every major candidate. It was by that time a conventional operation. Scores of national surveys were conducted every year sponsored by candidates, by the mass media, and by private interest groups. Polls were even being collected in a special library, The Roper Public Opinion Research Center in Williamstown, Massachusetts, where most of the national polling organizations deposit their old IBM cards. Polling had come of age.

In 1960 with the aid of the public opinion poll data thus being accumulated, and with the use of computers, a new research technique—simulation—came onto the scene. A description of its first and primitive political use in 1960 is the subject of this monograph. How soon and how fully this new instrument will become assimilated into the normal practice of politicians and of political scientists time alone will tell, but the limited experience of the first experiment already limns broadly a visage of the future.

What Simulation Is

Before we start our story it would perhaps be well to define and describe what computer simulation is.

Purpose

Simulation is a technique that is appropriate when so many variables are simultaneously in operation that simpler methods of calculation fail. It is a technique used when the equations that describe a system involve stochastic variables, or discontinuous variables, or such a complex of variables as to make an analytic solution impractical or impossible. A

2

computer can count, compare, list, add, subtract, multiply, divide. A computer performs no operation that the analyst who has programed it cannot also perform. Its merit is that it does these things fast, tirelessly, without forgetting its preliminary results.

Procedure

In a computer simulation the entities studied (and also their attributes) are represented by symbols in computer registers. These symbols are changed step by step to represent expected changes in real-world entities as known variables act upon them. The transformation rules that specify which symbols are to change, when, and how, are contained in a set of computer instructions called "the program." Thus, simulation is a way of making a computer act out a history of expectable processes that could occur in a complex real-world system.

Types of Simulation

The transformation rules embodied in the simulation program may be *stochastic*. That is to say, instead of asserting that "Under condition X, A changes to B,"[2] the program may assert that "Under condition X, A changes to B, with a probability of p." If the computer finds condition X, it goes through a so-called "Monte Carlo routine." Figuratively, it spins a roulette wheel, and, if the outcome is from 0 to p, A changes to B. If the outcome is from p to 1.0, it does not. Simulations with such routines are called stochastic. Simulations with no Monte Carlo processes are called deterministic. The election simulation we are about to describe is deterministic.[3]

[2] To put it more accurately: in the real world, under condition X, A changes to B. Parallelly, in the program, whenever symbols X' are found in the computer memory, then symbol A' is changed to symbol B'.

[3] In some usages the word simulation is reserved for stochastic simulations. Obviously that is not our usage here.

Many other distinctions among various types of simulations may be drawn,[4] for example the distinction between a simulation of a system versus a simulation of the results or output of a system.[5] (The simulation to be described emphasizes the output of a political system, that is, mass votes, without going very far into the sociopsychological detail of the processes by which individual vote changes come about).

When to Use Simulation

1. Simulation is a way of using scientific theories, not a substitute for them. To write the program for a computer simulation one must have some theories about how things change under various circumstances. Some substantial hypotheses must exist about the behavior of the entities being simulated as a first condition.

2. To represent a system of entities and their attributes, data describing the initial state of the system is necessary. In simulation jargon, one needs to have values for the parameters of the system. Some of these parametric values

[4] A good idea of the great variety of conceptions of simulation within the behavioral sciences may be gleaned from six recent volumes: Harold Borko (Ed.), *Computer Applications in the Behavioral Sciences* (Englewood Cliffs, N.J.: Prentice-Hall, Inc., 1962); Edward Feigenbaum and Julian Feldman (Eds.), *Computers and Thought* (New York: McGraw-Hill Book Co., 1963); Harold Guetzkow (Ed.), *Simulation in Social Science: Readings* (Englewood Cliffs, N.J.: Prentice-Hall, Inc., 1962); Edward Holland with R. W. Gillespie, *Experiments on a Simulated Underdeveloped Economy* (Cambridge, Mass.: M.I.T. Press, 1963); Guy H. Orcutt, Martin Greenberger, J. Korbel, and A. Rivlin, *Microanalysis of Socioeconomic Systems: A Simulation Study* (New York: Harper & Row, Publishers, Inc., 1961); Silvan Tomkins and Samuel Messick, *Computer Simulation of Personality* (New York: John Wiley & Sons, Inc., 1963). An excellent technical introduction to computer methods in behavioral science has been given by Bert F. Green, Jr., *Digital Computers in Research: An Introduction for Behavioral and Social Scientists* (New York: McGraw-Hill Book Co., 1963).

[5] Robert P. Abelson, "The Use of Surveys in Simulations," *Public Opinion Quarterly*, Vol. 26 (1962), pp. 485–486.

will come from measurement; that is, they are actual data. (In our study the data were mostly public opinion poll results.) Other parameter values may be guesses. One way to use computer simulation is to make several different guesses about a parameter and then run the simulation with each of the different settings in turn, to see what difference the value of the parameter makes. That procedure is known as *sensitivity testing*. However, to use simulation for prediction of real-world events, one cannot rely completely on guessed parameters. So a second condition for predictive use of computer simulation is that there be some data to start with.[6]

3. Having theories and data are not sufficient conditions to justify using computer simulation. A third condition is essential, namely that the calculations be so numerous as to make a pencil and paper solution an onerous chore. One uses computers when the transformations which the program makes on the data are tediously iterative. (In the simulation we are about to describe, we examined the behavior of each of more than 100 different types of voters in each of 48 states separately before putting them back together to make election prognoses.)

A tangible experience of what is defined makes a definition meaningful. The description of the simulation of the 1960 political campaign which follows may give life to the foregoing abstractions.

The Origins of the Idea

In 1959, when the idea of an election simulation was first broached, the art of computer simulation was but ten years old. It was an art mostly practiced by engineers. The first simulations dealt with military problems.

---

[6] In the simulation we are about to describe, while the data that gave parameter values were mostly derived from public opinion polls, there were also some guessed parameter values; the sensitivity of the system to some of them will be tested in Chapter 2.

A classical example of a simulation problem is tactical maneuvering in a tank or naval battle. Symbols standing for tanks or ships are projected on a map. These are moved around the map according to programs expressing tactical doctrine. They "shoot" from time to time in accordance with the rules. Each shot is evaluated for its probable outcome given the terrain in which it occurs, the relative positions of the combatants, and the nature of their equipment. Each mock engagement has an outcome. A series of them gives some indication of the statistical chances of success under alternative strategies.

Such simulations were familiar in 1959. So were simulations of engineering designs: for example, a simulation of the stresses and strains on an as yet unbuilt bridge, or of the flight characteristics of a new aircraft. And so were simulations of a job shop which were used to spot where bottlenecks and queues would develop as work flowed from one machine to another under different work assignment rules. Another important simulation was that by Newell, Shaw, and Simon,[7] in the imitation of individual logical thought processes. Under the name of "artificial intelligence," a computer had been programed to prove theorems in geometry.[8] Alex Bernstein[9] and Arthur Samuel[10] had programed computers to play chess and checkers.

[7] Allen Newell, J. C. Shaw, and H. A. Simon, "Elements of a Theory of Human Problem Solving," *Psychological Review,* Vol. 65 (1958), pp. 151–166.

[8] H. Gelernter and N. Rochester, "Intelligent Behavior in Problem-Solving Machines," *IBM Journal of Research and Development,* Vol. 2 (1958), pp. 336–345.

[9] Alex Bernstein and Michael de V. Roberts, "Computer versus Chess-Player," *Scientific American,* Vol. 198 (1958), pp. 96–105.

[10] Arthur Samuel, "Some Studies in Machine Learning Using the Game of Checkers," *IBM Journal of Research and Development,* Vol. 3 (1959), pp. 211–229.

All of this work was brand new, and almost none of it was in the social sciences. There existed as yet no notable published examples of social simulations: that is, simulations in which

a. some of the entities symbolized human beings, and
b. the programs expressed behavioral science propositions on how persons interact.

But if the social sciences were lagging behind in discovering uses of simulation, they were not far behind. The idea of social simulation was in the air and a few pioneers had already started work, most notably, William McPhee and James Coleman.[11] Early in 1959 McPhee had already developed a simulation of electoral behavior, which one day he described to a friend, Edward Greenfield, and at his suggestion to Ithiel Pool. It was a model designed to trace long political waves in the four-year election cycle. These waves occur as new age groups enter the voting population during eras of different political climate. At the same time older voting groups are dying out. McPhee and Pool agreed that this model, useful as it might be for its purposes, was not the one that would answer the action questions that a candidate might ask during a campaign.[12] They agreed, however, that such a model could

[11] Much of McPhee's early work has since been published in his *Formal Theories of Mass Behavior* (New York: The Free Press of Glencoe, 1963).

Cf. James Coleman, *Introduction to Mathematical Sociology* (New York: The Free Press of Glencoe, 1964), and "Analysis of Social Structures and Simulation of Social Processes with Electronic Computers," *Educational and Psychological Measurement*, Vol. 21 (1961), pp. 203–218.

[12] It should be noted, however, that in subsequent months the McPhee model was adopted by IBM and CBS and totally reinterpreted. The time periods were scaled down from years to weeks and the simulation was used with substantial success to analyze the Wisconsin primaries. (See William McPhee, *op. cit.*, Chapter 4.)

be built. It would focus on items about which

a. a candidate has genuine strategic alternatives,
b. the behavioral sciences have some relevant theories, and
c. the necessary data existed.

A topic which met those criteria was that of campaign issues. To a degree, a candidate can choose the issues on which he stands. He chooses them in the speeches he makes. He debates every day with his close advisors how to use the media he can command. Should he in the next address speak out most loudly on civil rights, or talk instead about communism and national defense, or about conservation and natural resources? Candidates do inject the issues that distinguish campaigns. Eisenhower in 1952 promised to stop the war in Korea. Stevenson in 1956 introduced the issues of the draft and nuclear fallout. Kennedy in 1960 chose to popularize the issue of Medicare. In each case some advisors would have preferred that these issues be avoided.

Issues are not the most important factor in a campaign. Lazarsfeld and others have shown that social milieu and party are much more important than issues in determining a voter's decision.[13] The candidate's

---

James Coleman also developed a campaign simulation which he used in Baltimore. (James Coleman and Frank Waldorf, "Study of a Voting System with Computer Techniques," unpublished.)

Both the McPhee and Coleman simulations were designed to be used in conjunction with a special field survey. They were designed to explore the processes of voter decision making. They would be of more interest than our simulation as a social psychologist. They were not, however, designed for action research by a national party.

[13] Bernard Berelson, Paul F. Lazarsfeld, and William N. McPhee, *Voting* (Chicago: University of Chicago Press, 1954). Cf. also Angus Campbell, Philip E. Converse, Warren E. Miller, and Donald E. Stokes, *The American Voter* (New York: John Wiley & Sons, Inc., 1960).

image and personality may also be more important in determining the election outcome. But these are things about which the candidate can do little. He controls the issues he talks about. He has much less control of who he is and who the voters are.

Issues and how the voters respond to them are also topics that social scientists know something about. Both theories and data about them exist. The data are in public opinion poll archives. The standard public opinion poll questions have concerned attitudes on issues. We ask citizens, for example, whether they are for or against foreign aid, for or against civil rights legislation, whether they object to having a Catholic as President, and what they think are the most important issues. We are much less skilled at using surveys to learn whether our citizens respond well or poorly to the pancake make-up on a candidate's cheek, the inflections in his voice, or other personal traits. Nor are we very skilled at identifying what it is that makes an effective campaign ad or TV format. The candidate who asks what useful social science data are available to him had best be told about poll results on the public's attitudes on issues.

Moreover the behavioral sciences contain theories that bear upon the manner in which people's attitudes toward issues relate to how they vote. These theories go under various names—balance, congruity, cognitive dissonance, cross pressure. But whatever the language used, the theories deal with a human tendency to maintain a subjective consistency in orientation to the world.

The theory of cross pressures was formulated by Lazarsfeld, Berelson, and others in studies of the election campaigns of 1940 and 1948. The theory starts with the familiar observation that party affiliation is statistically related to certain social variables. In the

North rural people, upper-income people, Protestants, and (at that period) older people tend to be Republicans. Urban people, poor people, Catholics, Jews, and (at that period) Negroes and young people tend to be Democrats. In the South the correlations are different. Being rural and thus to a greater extent part of the old South increases the chances that one is a Democrat; Southern Republicans are found mostly in the cities.

Such correlations with social variables arise in various ways. In part they are reflections of tradition. Negroes became Republicans after the Civil War. It took the New Deal to obliterate that heritage in the minds of many Negroes. But tradition is not just a matter of an individual's habit. It is even more a matter of group reinforcement. People's politics are strongly shaped by the personal influence of those around them. Berelson, Lazarsfeld, and McPhee found that only 33 per cent of Democrats in Elmira, New York, said that among their three closest friends there was a Republican; only 20 per cent of Republicans said that among their three closest friends was a Democrat.[14] In each case the voter's image of the world was an overwhelmingly balanced one: sensible people would, of course, be for their preferred candidate and, naturally, the people they liked were such sensible people. This balanced view of the situation could arise in one of four ways. First, it is often the case that similar types of people tend to vote the same way for historical reasons or by virtue of group self-interest, and also find themselves in close social contact with one another. Thus small-town people know small-town people and tend as a group to vote Republican. Rich people know rich people, and tend to vote Republican. Catholics know Catholics and tend to vote Democratic. And so on.

[14] Berelson, Lazarsfeld, and McPhee, *op. cit.*, p. 81.

However, when social milieu and political views are out of balance, then further adjustment processes may be set in motion. The voter may compromise his views toward those of his friends, or attempt to convince his friends of the merit of his own views. Failing this, he may manage to misperceive his friends' views to produce the illusion of consensus. Some people undoubtedly find it hard to believe or admit that an admired friend could be so foolish as to favor the other party. Lastly, if a person finds that his friends' political (and other) views are uncongenial, he may try to find new friends whose views are more congenial.

Of the three adjustive processes—social influence, misperception, and social locomotion—it seems likely that the first is most common when a few people with strong views interact with many people with weak views; the second is most likely when people with moderately strong views interact socially but do not often discuss politics and are thus unaware of one another's real positions; and the third is most likely when people with moderately strong views interact socially and discuss politics, but find themselves within a fluid enough social network to change their friendship circles.[15] (Of course if friendship circles are based upon membership in social categories, as they often are, social locomotion is not easy: the poor man cannot freely associate with rich men, nor can a person readily change his religion, color, or ethnic origin. In cases of social entrapment in a group with unambiguously uncongenial political views, a person is likely to withdraw from political discussion or from an interest in politics altogether.)

---

[15] All three of these balancing mechanisms, but most especially friendship changes, were observed in a recent study of experimentally created social groups by T. N. Newcomb, *The Acquaintance Process* (New York: Holt, Rinehart, and Winston, Inc., 1961).

There is, in any case, a minority for whom this consistency of influence does not prevail. Such people are said to be under cross pressure. The lifelong Democrat who is a rich, rural, Protestant is under cross pressure. So is the rich urban Catholic. The latter's coreligionists in the city mostly press him toward what was the group's traditional Democratic affiliation. His wealthy business colleagues press him to a Republican one.

Broadening the concept somewhat, we can also think about cross pressures arising from issue-attitudes. The Vermont farmer who is approaching 65 and strongly in favor of Medicare may find himself torn between normal Republican affiliations and a Democratic issue stand. A big-city industrial laborer who thinks domestic Communists have infiltrated the government and who does not want Negroes living near him may find himself under cross pressure between his labor interests and his other views. Cross pressure may also characterize the man who wants both a balanced budget and more national defense. Cross pressures come in many forms.

What do we know about men under cross pressure? How do their vote decisions differ from those of persons on whom pressures are consistent? The Lazarsfeld-Berelson studies present several hypotheses.

1. It is among the people under cross pressure that we find the individuals who change their minds from election to election and within the campaign period.

2. However, even among these people most end up voting for their traditional party in November. While some resolve the conflict by switching, more of them resolve it by such other means as misperceiving where their party stands (that is, a Democrat may accept as fact his platform's ceremonial tributes to balanced budgets).

3. Some people under cross pressure resolve their conflict by withdrawing cathexis from politics. Men under

cross pressure are less interested in the campaign, and a higher proportion of them become nonvoters.[16]

4. People under cross pressure make up their minds late in the campaign period. People under unified pressures are apt to have their minds made up from the beginning.

All of these statements about election campaigns are consistent with the more general formulations of balance theory.[17] Balance theory is a general formulation of some aspects of the psychology of attitude formation and cognition. Let us consider balance among three objects, one of them a cognizing human being whom we will call "*P*" and the other two the objects *X* and *Y* of which he is cognizant. *P* can feel positively or negatively or neutrally toward *X* and toward *Y*. He can also believe assertions about the relation of *X* and *Y*. An assertion may say that *X* is positive or negative or neutral toward *Y*. Suppose that *P* feels positively toward *X* and *Y*; the situation would be balanced only if he then also believes *X* and *Y* are positive toward each other. For example, the statement "my friend favors the Democratic candidate" is a statement which, if said by a Democrat, asserts a positive relationship between two positive objects, "my friend" and "the Democratic candidate." That is a balanced situation. So is the statement "my candidate opposes corruption in government." On the other hand for a Republican to say "my friend favors the Democratic candidate" represents an

---

[16] We will note in Chapter 2 some reasons for qualifying this hypothesis. It is, however, stated by Berelson, Lazarsfeld, and McPhee, *op. cit.*

[17] Cf. Roger Brown, "Models of Attitude Change," in R. Brown, E. Galanter, E. H. Haas, George Mandler, *New Directions in Psychology* (New York: Holt, Rinehart, and Winston, Inc., 1962); Robert P. Abelson, Milton J. Rosenberg, "Symbolic Psycho-Logic: A Model of Attitudinal Cognition," *Behavioral Science,* Vol. 3 (1958), pp. 1–13.

imbalance. A positively valued object is asserted to be positive towards a negatively valued object.

The prediction of balance theory is that when imbalance exists there is some tendency to restore balance or at least remove imbalance. This may happen by a change in some belief (for example, switching by persons under cross pressure as in the above hypothesis 1 about voting), or by misperceptions (for example, hypothesis 2), or by withdrawing attention from the subject (for example, hypothesis 3), or in other ways.

The theories of cross pressure and balance offer a body of behavioral science hypotheses with action implications for politicians. They were a natural basis for an election simulation.

As we have already indicated, however, theories alone without data will not permit a predictive simulation in a world of transiency and change. Essential to our entire operation was the existence by 1960 of a vast file of old public opinion survey cards which could serve to provide measurements of the attitudes of different groups in the population toward relevant issues. The study on which we embarked was a "secondary analysis" of old poll results. Students of public opinion are becoming aware that the growing backlog of earlier polls provides a powerful tool to aid in the interpretation of new poll results. While polling has now been routine for three decades, poll archives are just beginning to be assembled. The main one is the Roper Public Opinion Research Center in Williamstown, the existence of which made feasible the project described here.

The first step in the project was to identify in that archive all polls anticipating the elections of 1952, 1954, 1956, and 1958. (Pre-election polls on the 1960 contest were added later when they became available.) We selected those polls that contained standard identification data on region, city size, sex, race, socioeconomic status, party, and religion. The last item was the one most often missing. Further, we restricted our attention to those polls that asked about vote intention and also about a substantial number of preselected issues such as civil rights, foreign affairs, and social legislation. From 1952 to 1958 we found 50 usable surveys covering 100,000 respondents. Fifteen polls anticipating the 1960 elections were added to this number. The 66 surveys represented a total of over 130,000 interviews.

The organization of these data was a substantial year-long chore. It could not have been done before the advent of the electronic computer and of magnetic tape. The Presidential election of 1960 was the first in which all the technological prerequisites for our project existed: survey archives, readily available tape-using large-memory computers, and previously developed theories of voter decision.

THE POLITICAL HISTORY
OF THE PROJECT

A key figure in the initiation and carrying out of the project was Mr. Edward Greenfield, a New York businessman actively engaged in Democratic politics. Through him a group of New York reform Democrats who had taken major responsibility for raising money for the Democratic Advisory Council became interested. They carried the project to the Advisory Council. Before this group of individuals was willing to secure funds they wanted to be sure that the results were likely to be scientifically valid and practically useful to the Party. In May of 1959 the project was discussed in Washington at a meeting attended by Mr. Charles

Tyroler, Executive Secretary of the Democratic Advisory Council; the members of the Council executive committee; Paul Butler, Chairman of the Democratic National Committee; several other officials of that Committee; Mr. Neil Staebler, Michigan State Chairman; and a number of social science consultants, including Samuel Eldersveld, Morris Janowitz, and Robert Lane. This group was interested but reserved. It was suggested that the project should be supported for four months initially and at the end of this period a further review should be made.

The Williamstown Public Opinion Research Center agreed to permit the use of polls in their archives on two conditions: first, all basic data tabulated from their cards were to be made available to the Center so the Republican Party would have an equal opportunity to use them if they wished; second, (requested also by the social scientists engaged in the study) all results could be published for scientific purposes after the election. Both these conditions were met: the print-out of the data on the computer tape was supplied to the Center (not the simulation programs, nor supplementary data obtained from other sources such as the census) and this report is now published.

The principals organized themselves as The Simulmatics Corporation, with Mr. Greenfield as President; he managed the project throughout. Although the project was fundamental scientific research, it was clear that universities would not and should not accept financing from politically motivated sources or permit a university project to play an active role in supplying campaign advice to one party. The project required a private vehicle. The Simulmatics Corporation has continued since as a commercial company doing applied social simulations under the guidance of Mr. Greenfield. In 1960, however, it was not yet a conventional company; at that time it was a single-purpose organization to do a political job.

The summer of 1959 was devoted to data reduction. In October 1959, when the preliminary data processing had been substantially completed, a review meeting in New York was attended by many of the same persons who had been at the Advisory Council meeting in May, plus a number of social science consultants. Included among the advisors were Harold Lasswell, Paul Lazarsfeld, Morris Janowitz, and John Tukey. Although the degree of confidence in the basic approach ranged from enthusiasm to doubt, a decision to proceed was quickly reached.

The next step was the development of computer programs to perform the various operations which are described in the subsequent sections of this chapter. One objective was to make possible rapid inclusion of new data which might, we hoped, become available during the campaign. Our hope was only slightly fulfilled.

By June of 1960 a report on the Negro vote in the North was prepared. It was a sample of what might be done by the Simulmatics process.

Our contractual arrangements with our sponsors ended with the preparation of the process and of this report illustrating it. Any further reports on specific topics were to be purchased by the Party in the precampaign and campaign period at their discretion. In the immediate preconvention period, however, the National Committee felt that it should not make decisions that would shortly be the business of the nominee. After the convention, the Kennedy organization, contrary to the image created by the press, did not enter the campaign as a well-oiled machine with a well-planned strategy. Except for the registration drive, which had been carefully prepared by Lawrence O'Brien, no strategic or organizational plan existed the day after the nomination. It took until August for the organization to shake down. No campaign research of any significant sort was therefore done in the two months from mid-June to mid-August, either by Simulmatics or by others. In August, a decision was made to ask Louis Harris to make 30 state surveys for the Kennedy campaign. However, because of the late start, data from these surveys would not be available until after Labor Day. On August 11, the Kennedy campaign management asked The Simulmatics Corporation to prepare three reports: on the image of Kennedy, the image of Nixon, and foreign policy as a campaign issue. These three reports were to be delivered in two weeks for use in campaign planning. Along with them we were to conduct a national sample survey which, in the minds of the political decision makers, would serve to bring the Simulmatics data, based as they were on old polls, up to date. (It should be mentioned that one of the most difficult tasks of the Simulmatics project was persuading campaign strategists that data other than current intelligence could be useful to them.) The national survey by telephone was conducted for the project by the Furst Survey Research Center. It was extremely useful in guiding the use of the earlier data. It confirmed the published Gallup finding that Nixon was at that point well in the lead, though we disagreed on the proportion of undecided voters (we found 23 per cent). It made us aware that women were responsible for Nixon's lead. It also persuaded us that voters were largely focusing upon foreign policy at that point in the campaign.

The relationship between such current intelligence and a simulation model developed out of historical data is analogous to the relationship between current weather information and a climatological model. One can predict tomorrow's weather best if one has not only current information but also historical information about patterns into which current reports can be fitted. While it would be presumptuous to assert that in two weeks of intense activity we approached an effective integration of the two sets of data, that was the ideal we had in mind and which in some limited respects we approximated.

It should be added that the introduction of the national survey results was possible only because of prior preparation for rapid data analysis. The survey was ordered on a Thursday, the field interviewing took place between Saturday and the following Thursday, by Friday morning all cards had been punched, and by Friday night the preprogramed analysis had been run and preliminary results were given to the National Committee.

The three reports that had been ordered on August 11 were delivered on August 25. The swiftness of the entire operation is, of course, a testimony to the advantages of a high-speed computer system. Nonetheless, such intense pressure is not an optimum condition for research work, even though rapid analysis was one of our objectives from the start. The reader who suspects that under those circumstances clerical errors inevitably occurred is quite right. It was our good fortune that none of those which we have found since in rechecking have caused us to alter any conclusions. But such tight schedules are not a happy normal mode of research work. Nevertheless the possibility of such turnabout times in well-prepared computer operations opens up new ranges of opportunities for action research in the social sciences.

The reader may ask whether the large preparatory effort was justified in terms of the quantitatively limited use of the project. When we planned the project, we anticipated—perhaps unrealistically—active campaign work from the beginning of the summer until about September 15. (Anything done later than that would be of far less use.) How far the investment was justified by the two weeks of work actually done is a question that we find impossible to assess. The answer depends on an estimate of how much impact the contents of the reports had on the campaign. These received an extremely limited elite circulation. They were seen during the campaign by perhaps a dozen to fifteen key decision makers, but they were read intelligently by these talented and literate men.

The late President Kennedy was an avid and able user of political research. The same can indeed be said of the whole generation of politicians "born in this century" who appeared on the national scene at the beginning of the sixties. It had not been true of any national leader before them. President Truman never liked the pollers and has lambasted them often with juicy epithets. Eisenhower neither used nor understood them well. Neither did Stevenson, who read research reports but tended to act in utter disregard of their findings.

The same thing cannot be said of any of the top political figures of 1960. Nixon, Rockefeller, and Kennedy all relied on polls, read them carefully, and understood them. A new political generation finally completed the revolution that had begun with Gallup's technological innovation in 1936. John F. Kennedy in particular not only understood enough to trust research; he understood enough to know when and in what respects to distrust it. He could ask the right questions and could distinguish between findings and implications. The same capacity for remembering and using numbers that so awed economists who dealt with him also stood him in good stead with survey researchers.

The question most often asked us by friends and readers, as if we could answer it, is how influential were our reports? (Some popular political reporting and a work of fiction has attributed to them an incredible influence.) The belief that an answer to this question can be given with any precision is based upon a misperception of the policy-decision process. Neither we, nor the users, nor even John F. Kennedy if he were alive, could give a certain answer.

When a policy maker reaches a decision, he knows what conscious factors enter into it. But he seldom knows which conversations that he has had or words that he has read were responsible for initially injecting those ideas into his head, or reinforcing them, or turning them into final convictions. It is rare indeed that any idea comes from only one source. In a Presidential campaign there are only a limited number of policy alternatives, and there are myriad voices arguing for each of them. Our own contribution, if any, was to bolster by evidence one set of alternatives. With one exception they were the alternatives that the candidate ended up choosing. We flatter ourselves to believe that that fact showed his sagacity; we know it was not due to a simple cause-effect relationship between our data and his action.

Specifically, our main contributions concerned the religious issue, civil rights, and stress on foreign policy.

Data and arguments on each of these issues were coming from many sources, among them the experienced "pols," newspapermen, the special interest groups, as well as researchers. Gallup's and other polls were read avidly. The main research advisor of the Kennedys was Louis Harris, who at that time had a personal relationship with them and saw them frequently during the campaign. His operation and ours were entirely independent. We did not use his polls which were not nationwide, and which, except for the preprimary surveys, especially in Wisconsin and West Virginia, did not become available until well along in September, after which they were very useful for quick current strategy appraisals. His advice based on current hot polls and ours based on simulation from the past tended mostly to coincide. On the religious issue both of us agreed with those who recommended frankness and directness rather than avoidance.

But certainly when Kennedy decided to confront the bigots head on, he himself could not say what part in his decision was played by any one piece of evidence. His own impulses, his sense of the morally right thing to do, his politician's sense that people would respond favorably to the right action, the urgings of troubled fellow-Catholics around him, the advice of some shrewd politicians were simply sustained by research results that suggested that the same course would be beneficial. Certainly no one can say how much pause it might have given John F. Kennedy if the research results had coincided with the views of the equally shrewd advisors on the opposite side of the issue.

Leaving aside as unanswerable the question of how much influence any given information had on anybody's mind, we can state only a boundary condition of influence—who had the information available, and when. The first report on "The Negro Vote in Northern Cities" (which was a use of our data bank rather than strictly a simulation) was given to Chester Bowles, Chairman of the Democratic Party Platform Committee, and to the member of Senator Kennedy's staff who did the main drafting of the civil rights plank of the platform, Harris Wofford, in June shortly before the platform drafting. (Certainly it would be silly to think that the evidence there presented on the political importance of winning this vote *caused* the strong civil rights convictions of these two dedicated men!) The second, third, and fourth reports were, as noted, turned in just before Labor Day. They went to Robert Kennedy. When they were presented, there was also a briefing session with him and his top staff. Primary interest at that time in the campaign was focused on the religious issue and how to handle it, on the dramatic slippage of Kennedy behind Nixon attested to by the Gallup poll and confirmed by ours, and on the lag of Kennedy's popularity among women.

Foreign policy was the issue on which Kennedy's campaign strategy deviated somewhat from that suggested by our data. We will consider that, as well as the civil rights and religious issues, below. Suffice it at this point to note that our analysis would suggest that Kennedy's heavy emphasis on foreign rather than domestic policy in the last ten days of the campaign was disadvantageous to him.

Let us summarize our reply to the inevitable question about the use of our research:

Despite the contraction of our effort below our original hopes, our feeling is one of relative satisfaction that the Simulmatics project was able to provide on demand research concerning the key issues at perhaps the critical moment of the campaign. While the campaign strategy conformed rather closely, except on a few points, to the advice in the more than one hundred pages of the three reports, we know full well that this was not because of the reports. Others besides ourselves had similar ideas. Yet, if the reports strengthened sound decisions on a few critical items, we would consider the investment justified.

ORGANIZING THE DATA

The data on which our research was based consisted of questionnaires and code books from 50 (later 65) national surveys and reply punch cards from 100,000 (later 130,000) individuals.

Such massive data required substantial innovations in analytic procedures. We were faced with a relatively new problem for public opinion researchers. The problem, by analogy, was how to do research on a library rather than on a single book. Survey analysts up to now have generally had to report and make sense of a single survey. Sometimes in trend studies or in cross-cultural or comparative research two or three or more surveys have been compared, but always in small enough numbers to treat each one as a single object of analysis. But in this study, to gain the advantages that could accrue from using 100,000 cases covering a decade of research, we had to find ways to simplify and to standardize what would otherwise have been an unmanageable mass. The study was, in one of its aspects, an experiment in methods for research in an emerging era of abundant libraries of survey data.

In essence, the data available to us were reduced to a 480×52 matrix. The number 480 represented types of voters, each one being defined by socioeconomic characteristics. One voter-type might be "Eastern, metropolitan, lower-income, white, Catholic, female Democrats." Another might be "Southern, rural, upper-income, white, Protestant, male Independents." Certain types with small numbers of respondents were reconsolidated, yielding the total of 480 types actually used.

The number 52 represented what we called in our private jargon "issue-clusters." Most of these were political issues, such as foreign aid, attitudes toward the United Nations, and McCarthyism. Other so-called issue-clusters included such familiar indicators of public opinion as "Which party is better for people like you?", vote intention, and nonvoting. In sum, the issue-clusters were political characteristics on which the voter-type would have a distribution.

One can picture the 480×52 matrix as containing four numbers in each cell. The first number stated the total number of persons within a voter-type who had been interrogated on that particular item of information. The other three numbers trichotomized those respondents into the percentages pro, anti, and undecided or confused on the issue.

We first assembled such a matrix for each biennial election separately. For example, to make up a 1952 matrix, we took all the selected surveys which occurred in the two-year period ending in November 1952. (Most of those surveys would have been in the four months before the election.) The number of surveys for each election period were:

1952	16
1954	15
1956	5

1958   14
1960   15 (containing only surveys up to January 1960)

The first four basic matrices were compiled by Spring 1960 and were cumulated together to form what we called a 4-period master matrix. Most of the work done by us during the 1960 campaign was done on the 4-period master matrix.

In addition, in the summer of 1960 we were receiving 1959 surveys and making up the 1960 matrix from them. The original idea was to work with a 5-period master and, if we had worked into September, we would have done so. But the programs for incorporating the new data were not all functioning in August. In the years following 1960 we have often used the 5-period master, and in the reports to follow it will be indicated in each case whether the 4-period or 5-period master is used.

Three questions arise about this operation: first, how did we select the surveys we used; second, how did we come to define the 480 voter-types in the way we did; third, how did we define the 52 issues.

We specified the following requirements for a survey:

1. It had to use a national cross-section or probability sample.
2. It had to be an election poll, asking the respondent how he intended to vote. It also had to have some additional political questions: how the individual had voted in the past, had he voted or did he intend to vote, and so on. Not all of the latter items were required on all surveys, but we were interested only in surveys with a certain minimum richness.
3. The surveys also had to contain questions about a

number of the issues we had listed as potentially important. While no survey contained information on more than a few such issues, we were interested only in surveys with questions on at least four or five issues.

4. The face-sheet data on the survey had to cover all of the items necessary to classify respondents into one or another of the 480 voter-types into which we were distributing respondents. The types are indicated in Table 1.1. It will be noted that we did not use all possible combinations of all variables which would have given us some 3,600 voter-types, many of which would have been infinitesimally small: for example, Southern, rural, Catholic, upper-socio-economic-status (SES), Negro, female Republican. Table 1.1 indicates the combinations that were actually used.

The regional division of states was

East: Maine, New Hampshire, Vermont, Massachusetts, Rhode Island, Connecticut, New York, New Jersey, Pennsylvania

South: Virginia, North Carolina, South Carolina, Georgia, Florida, Alabama, Mississippi, Arkansas, Louisiana, Oklahoma, Texas

Midwest: Ohio, Michigan, Indiana, Illinois, Wisconsin, Minnesota, Iowa, Missouri, North Dakota, South Dakota, Nebraska, Kansas

West: Montana, Arizona, Colorado, Idaho, Wyoming, Utah, Nevada, New Mexico, California, Oregon, Washington

Border: Maryland, Delaware, West Virginia, Kentucky, Tennessee

Since American survey organizations follow highly standardized procedures on face-sheet data, only a few of the criteria used in selecting surveys for the data bank restricted seriously the number of surveys that could be used. Socioeconomic status is not always measured in exactly comparable ways, though the bulk

27

TABLE 1.1
DEFINITION OF THE 480 VOTER-TYPES*

7 SES and City-Size Levels	Protestant Males			Protestant Females			Catholic Males			Catholic Females			Jews			Negroes		
	R	D	I	R	D	I	R	D	I	R	D	I	R	D	I	R	D	I
In each of 4 regions: South, East, Midwest, West																		
1. A Urban	R	D	I	R	D	I	R	D	I	R	D	I	R	D	I	R	D	I
2. A + B Town	R	D	I	R	D	I	R	D	I	R	D	I						
3. A + B Rural	R	D	I	R	D	I	R	D	I	R	D	I						
4. B Urban	R	D	I	R	D	I	R	D	I	R	D	I	R	D	I	R	D	I
5. C Urban	R	D	I	R	D	I	R	D	I	R	D	I						
6. C Town	R	D	I	R	D	I	R	D	I	R	D	I						
7. C Rural	R	D	I	R	D	I	R	D	I	R	D	I						

4 × 96 cells = 384 cells

7 SES and City-Size Levels	Protestant Males			Protestant Females			Catholic Males			Catholic Females			Jews			Negroes		
In 1 region: Border																		
1 + 2 + 3 combined	R	D	I	R	D	I	R	D	I	R	D	I	R	D	I	R	D	I
4 + 5 + 6 + 7 combined	R	D	I	R	D	I	R	D	I	R	D	I	R	D	I	R	D	I

1 × 36 cells = 36 cells

	Rep.	Dem.	Ind.
1. Protestants without SES rating in non-labor-force families	R	D	I
2. Catholics without SES rating in non-labor-force families	R	D	I
3. Persons of no religion	R	D	I
4. Persons of other religion	R	D	I

Add, in all 5 regions

5 ×

12 cells = 60 cells / 480 cells

* Key: R means Republican; D means Democratic; I means Independent.
A means professional, executive, and managerial occupations; B means white-collar occupations; C means blue-collar occupations.
Urban refers to metropolitan areas exceeding 100,000 population; town refers to communities between 5,000 and 100,000; rural refers to places under 5,000.

of surveys used a basically similar classification. Political party preference, not always asked, is virtually always asked on election surveys and therefore presented no problem. Regional classifications are quite standard. Rural-urban classifications tend to be standard at any given time, although over the decade there had been some changes corresponding to changes in census practice. The greatest difficulty arose with the religious notation. This is a highly sensitive item in the United States and on many surveys it is not asked. Since in 1960 it was obviously going to be a crucial variable, we were compelled to insist upon it and to exclude surveys on which the respondent's religion had not been asked.

If we had felt free to use one more variable, it would have been age. Age is related to voting behavior. Young voters behave differently from old, partly as a true function of age, but even more because they belong to a generation that grew up in a different era. But to have added three levels of age to our type criteria would have given us 1,440 types with many types having only a handful of cases to represent them.

A decision had to be made as to how many types we could effectively use and which variables were politically most predictive. We decided to have about 500 types, since with something like 100,000 cases divided into 500 types we would have an average of 200 cases per type. Attitudes on an issue considered in as few say as 10 out of 50 surveys would then be measured by an average of 40 cases per type. This being an average, the figure would fall well below that for some types. Clearly about 500 seemed a maximum number of types to work with.

Ranking political variables as to their significance caused us little doubt except for the variable age. Education, which at first glance might seem a grave omission, is highly enough correlated with socioeconomic status that it would have added little predictive power.

Since we were forced to drop age and education as type-defining variables, we treated them instead as issues 51 and 52.[18] That is to say we recorded for each of the 480 types the number of people in it who were young and who were old and the number with high and low education. It should be noted that although these issues were part of the data bank describing each of the 480 voter-types, they were never used.

Indeed it is perhaps well to point out that we use the word issue very loosely here to describe 52 sets of survey results. A list of the issue names can be found in Table 1.2. A glance at them reveals that most are issues in the usual sense and that is why we use the word. But a few of the so-called issues are sets of replies to survey questions dealing with such matters as past vote, nonvoting, and other behaviors not well described by the word issue. Since words are chosen for convenience and mean what we choose to make them mean, we shall continue to use the words issue or issue-cluster for the 52 sets of survey responses. The reader, however, should remain aware that when we talk of an issue we simply refer to some topic about which people were questioned in a number of different surveys.

The first step in organizing our data was to identify in each survey those questions which seemed to bear on any of the 52 issue-clusters we had listed as relevant to the campaign. But there are problems to this procedure.

Even if one finds numerous surveys, all of which deal with a common issue, they do not all ask the same

[18] In some earlier publications we talked of 50 issues, in others of 52. This is the basis of the discrepancy.

TABLE 1.2
THE FIFTY-TWO ISSUE-CLUSTERS

1. Attitude Toward Truman
2. Anti-McCarthyism
3. Attitude Toward Nixon
4. Anti-Nixon Voting
5. Attitude Toward Eisenhower
6. Attitude Toward Stevenson
7. Symington versus Nixon
8. Kennedy versus Nixon
9. Pro-Negotiationism
10. Defend Europe
11. Sympathetic to Europe
12. Neo-Fascist
13. Withdraw Troop Commitments
14. Anti-Catholic
15. Pro-Labor
16. New Deal Philosophy
17. Civil Rights
18. Anti-Administration
19. Congressional Votes
20. Which Party for Prosperity
21. Which Party for Personal Finances
22. Which Party Best in Crisis
23. Best Party for People Like Self
24. Normal Presidential Vote
25. Congress versus Ike
26. Stevenson versus Eisenhower (1952)
27. Stevenson versus Eisenhower (1956)
28. Union Membership
29. Fear of War
30. Concern with Rising Prices
31. Concern over Wages, Unemployment
32. Foreign Policy Knowledge
33. Salience of Civil Rights
34. Interest in Elections
35. Nonvoting
36. Extreme Nonvoting
37. Voting Intent Record
38. Ike Vote but Democratic for Congress
39. Which Party Best for Peace
40. World Situation Complacency
41. Meet with Soviets

(Continued)

TABLE 1.2 (*Continued*)

42. Red China Policy
43. Attitude Toward English, French
44. Attitude Toward Asian Neutrals
45. Attitude Toward Israel
46. Defense Commitments
47. Disarm H-Bomb
48. Preparedness
49. Attitude Toward U.N.
50. Attitude Toward Foreign Aid
51. Age
52. Education

questions. Let us take as an example the matter of civil rights. On one survey from the United States a question might be asked "Are you in favor of the Supreme Court's decision on school integration?" On another survey a question might be asked "Do you favor an FEPC (Fair Employment Practices Comission)?" On another the question might be "Should the schools in Little Rock be integrated now?" Each of these questions has one answer that can be characterized as pro-civil rights, another that can be characterized as anti-civil rights, and other answers that are indeterminate and nonresponsive. The answers can be recoded into such a trichotomy and then the cards from all the surveys thrown together into a single macrosurvey.

The reader will immediately note several difficulties of which the most prominent are (1) How can one know that different questions really deal with the same attitude? (2) Even if they do, don't they cut that latent attitude at different points on the scale?

In throwing several questions together as measures for a single attitude, one is in effect assuming them to be different manifestations of a single latent attitude. We know from much experience with factor analysis or scaling procedures that many verbal formulations do turn out to be explainable in terms of a much smaller number of fundamental attitudes. But how can we know which these are?

On a single survey we may establish the fact that two or more questions measure a single attitude by establishing a correlation between them across individuals. If almost all people who say aye to Question 1, say aye to Question 2, and if almost all people who say nay to Question 2, say nay to Question 1, then we believe that the two questions are measuring the same thing.

Between surveys we can do the same thing by correlations across groups. We cannot establish that the same individuals tend to answer two questions the same way since we have asked the questions of different individuals, but we can establish that subpopulations which tend to answer aye on Question 1 tend to answer aye on Question 2, and so on.

In doing the correlations we did not use the 480 voter-types because the number of individuals in a type who replied to any single question would have been much too small to permit the distribution of aye and nay replies to have been statistically stable. Indeed our purpose for combining questions into clusters of equivalent ones was to assure adequate numbers of cases of respondents in each of the 480 voter-types. At the state of the work before that combination was accomplished we were dealing with single questions from single surveys. For correlating such single questions across voter-types, we clearly had to use a restricted number of larger types to have cases enough in each type. We settled on 15 macro-types.

In the Simulmatics project we actually used a double-barreled criterion of latent attitude equivalence. First, questions were treated as dealing with the same topic only if we intuitively felt them to be so. Second, we required that each question distribute in the same way across the 15 macro-types before we would admit it to a cluster of equivalent questions.

It may be asked why we used a subjective criterion at all. We could have intercorrelated all questions with all other questions and by factor analysis or other procedure grouped those which intercorrelated highly. For certain purposes that would indeed have been the thing to do. However, for our purposes it seemed essential to define issues in ways close to the substance of current politics. It might be, for example, that responses to a question on school integration would correlate highly with responses to a question on atomic testing. This would prove some interesting things about the structure of political attitudes, but it would be hard to use a cluster containing such substantively diverse material in advising a candidate on the preparation of his speeches. Therefore we insisted both on substantive similarity subjectively measured and on correlations as criteria for question equivalence.[19]

[19] We should qualify this statement. What has been described is what we started out to do and what we did for most issue-clusters. In the end, however, we were forced to compromise on certain foreign-policy clusters. This in itself is an interesting finding. On almost all domestic questions, primarily because they were party-linked, it was possible to validate empirically the equivalence of questions which a priori seemed alike. On certain foreign-policy issues this was quite impossible. The political scientist looking at a half-dozen questions about foreign aid or about the United Nations might conclude that they all should reflect a common underlying attitude toward that matter. However, empirically, in many instances the distribution of replies was highly sensitive to conjunctural influences or shades in wording of the question. Rather than completely abandon the hope of doing any analysis of foreign-policy issues in the campaign, we retained some clusters which failed to meet the correlational test, labeling them *a priori* clusters, not sure of what we would do with them (in fact we did very little), but feeling it better to

Even when we have established that different questions are equivalent in the sense that they are tapping the same underlying attitudes, there remains the problem that the cutting points may be different. If the items form a Guttman scale, those who give an affirmative answer to one question may encompass all those who give an affirmative answer to a second question, but also some of those who say no on the second question. Fewer people in the United States favor an FEPC than are ready to accept school integration. Though both questions tap the same latent attitude, the latter does so in a milder form.

That is not, however, a real difficulty. If the sampling on two surveys is uniform, then the fact that different questions are used to tap a single attitude is no problem. To illustrate why the differences in cutting points is no great problem let us compare attitudes of two respondent types, I and II, the first of which is twice as numerous as the second. If we have drawn separate samples of the population in the same way, then, barring random variations, each sample will have twice as many persons of type I as of type II. Suppose on each sample we used a different measure of a single latent attitude, then our operational index of that latent attitude remains the proportion of affirmative replies even though these are replies to different questions. This is a legitimate procedure because each item in the mix of questions will be used with a constant proportion of each respondent type. Since we perform exactly the same complex operations on each respondent type, a comparison of the types by the results of these operations is legitimate.[20]

---

retain them on the computer tape than to discard the data from the start.

[20] More formally we throw m different survey samples together,

While such a combined measure is far more complex and less intuitive than a simple percentage of affirmative replies, it is logically acceptable. It is no different in principle from any index based on several questions occurring within a single questionnaire.

Conventional survey analysis is the case where the range of surveys being used (i) is 1; a single question on a single survey has been taken as the measure of an attitude. But clearly that is just a special case in which we are taking the average of a series of 1.

Whatever its logic, survey researchers are apt to find this innovation disturbing. But in fact they have been doing the same thing all along, not only in constructing indices on single questionnaires, but also in comparisons of surveys. No one has ever objected when, in comparative research, results from an identical question used on several different surveys have been compared. That is the heart of most time-series studies and most cross-cultural studies. Many researchers have shown that on some identical measure people felt different at time A than they did at time B, or in country A than in country B. In fact, however, the

---

each of size n_i, $N = \sum_{i=1}^{m} n_i$. If the sampling used in the m surveys is the same and the universe from which the sample is drawn is the same, then, barring random variations, there will be t respondent types in each survey and the proportion of respondents falling into any respondent type $P_i = n_{ij}/n_i$ is the same from survey to survey.

If different questions were used in different surveys to measure any issue attitude k, then the proportion among respondent type j answering the question that measures attitude k affirmatively $V_{ijk} = a_{ijk}/n_{ij}$ will vary between surveys. $a_{1jk}/n_{1j} \neq a_{2jk}/n_{2j} \neq \cdots \neq a_{mjk}/n_{mj}$. However, it was established by our correlational procedures for clustering different questions together to index an issue that the values of a_{jk}/n_j correlate across the t voter types for any given pair of surveys for any given issue k. Barring nonlinearities, random variations, and ceiling effects, for any two surveys, 1 and 2, for any two respondent types, 1 and 2, $V_{11k}/V_{12k} = V_{21k}/V_{22k}$

So in the combined sample of N respondents we construct a measure of attitudes on item k for use in comparing respondent types, which is an average across the m surveys.

measure is never identical. The same question couched in the same words, even if in the same language, does not mean the same thing at any two times or in any two cultures.

We had occasion to rediscover this point in the Simulmatics study. One of our issues was something called McCarthyism. It was the complex of attitudes toward domestic communism and was based upon a variety of questions such as "How well do you like McCarthy?", "Do you think Senator McCarthy has done more good than harm?", "Are the Democrats soft on communism?", and so forth. Some of these questions, particularly the question "How well do you like McCarthy?", were asked repeatedly over the period of our study.

When we listed those questions which seemed *a priori* to measure a single attitude, we of course included all of those where the wording was identical. But when we set up a correlation matrix to test our *a priori* grouping, we found that while various questions in the group correlated very well with each other, the question "How well do you like McCarthy?" did not correlate with the identical question asked at different times. On closer examination it became clear that the question had changed its entire meaning at the time that the United States Senate censored him. The distribution of attitudes toward McCarthy across voter-types before the censure and after the censure was different. What had earlier been a simple measure of anticommunism became after the censure also a measure of confidence in the United States Senate and its wisdom. Questions with other wordings were more equivalent to this question before the censure than was the question to itself before and after. Someone using traditional methods of survey research would treat as equivalent the verbally identical question as if that raised no difficulty, but would have been more hesitant about treating as equivalent the verbally different ones. However, it should be clear that it is just as legitimate to use verbally different questions from different surveys as to use verbally identical ones, and it is certainly more justifiable to use questions differently worded which are testedly equivalent than to use a verbally identical question both before and after its meaning had changed.

The reorganization of data we have just described, which took about a year because we were feeling our way in new and unexplored terrain, could now be done much more rapidly though it still would be a large operation. The end product was a set of matrices of attitudes on issue-clusters by voter-types: 480×52 matrices.

The data format and its transfer to high-speed tape facilitated its use in computer simulation of the effects of hypothetical campaign strategies.

FORTY-EIGHT SYNTHETIC STATES WITHOUT LOCAL POLITICS

In the data bank which we have just described, the five geographic regions of the United States were one of the variables by which voter-types were defined. American politics are notably regional. But in some respects they are even more localized than that.

Presidential elections are determined by states. It is not enough to guess the Democratic percentage in the Northeast; one needs to estimate which way New York will go with its 45 electoral votes. A favorite game all through a campaign is adding up electoral college estimates state by state.

One of the benefits gained from the large number of interviews we used was the possibility of approximating state-by-state results. A single national sample survey—even a relatively large one—has too few cases from most states to permit any significant analysis of state politics. The same would have been true, however, even for our voluminous data if we had attempted to do a state-by-state analysis in a simple way. We had an average of about 2,000 interviews per state, but that is a misleading figure. In a small state there might have been no more than 300 or 400 interviews. On a particular issue-cluster that had occurred in only one tenth of the surveys, for example, there would be too few cases for effective analysis. We therefore developed a system for creating synthetic states.

By an elaborate analysis of census, poll, and voting data—made more difficult because 1960 census results were not yet available—we developed a set of estimates on the number of persons of each voter-type in each state. (Note that since *region* was one of the defining characteristics for the 480 voter-types, there were at most only 108 voter-types found in any given state.) It was assumed that a voter of a given voter-type would be identical regardless of the state from which he came. A synthetic state therefore consisted of a weighted average of the voter-types in that state, the weighting being proportional to the numbers of such persons in that state.[21] For example, we assumed that

[21] In addition to the shortage of cases for direct state-by-state classification of respondents, a more sophisticated but equally compelling reason militates in favor of using a system of synthetic states. The survey organizations whose data we were using do not try to sample each state representatively. They could not do so within the confines of single surveys. They do try to sample regions representatively. But that goal can be achieved with great and continuing distortions of individual states. If for example one survey organization has a

the difference between Maine and New York is not truly a difference between New Yorkers and inhabitants of Maine as such, but a difference in the proportions of different voter-types which make up each state. We assumed that an "upper-income, Protestant, Republican, rural, white male" was the same in either state, and that a "small-city, Catholic Democratic, lower-income, female" was also the same in either. This assumption enabled us to use all cases of a voter-type from a particular region in arriving at a conclusion for a state.

We do not assert that the assumptions on which this synthesis is based are true. On the contrary we can be sure that they are partly false. The interesting question intellectually is how good were the results obtained with these partially true assumptions. The test is, of course, how far state-by-state predictions made on these assumptions turn out to correspond to reality. To the extent that they do, it suggests that the essential differences between states in a region are in distributions of types rather than in other geographic differences such as political history.[22]

---

rural sampling point for New England in Maine, with a good interviewer there, and has an urban sampling point with an interviewer in New Hampshire, then Maine will regularly have rural results and New Hampshire regularly urban and both will be distorted as states, without New England being distorted. For this reason it is not possible to divide good national or regional samples directly into good state samples.

Hawaii and Alaska were omitted because they had not been states and were therefore not included in surveys in the decade in which our data were originally collected.

[22] Clearly the accuracy of syntheses of states is critically dependent upon appropriate regional definitions. The states where the simulation was most notably off included Arizona, Nevada, New Mexico, Idaho, and Colorado, states mostly of small population, and states which, in the absence of a "Mountain Region" in our classification, we

In the remainder of this chapter we test these and other assumptions by seeing how closely the outcome of a simulation using them can correspond to the real-world outcome in November 1960. Later in this monograph we examine some simulations of what might have happened under various alternative scenarios of history—what might have happened if world news had been different, what might have happened if the candidates had conducted themselves differently. But first we look at simulations close to reality. They predict the state-by-state outcome rather well. The particular history of state machines and the details of political and organizational history, in which politicians and political scientists revel, meant little except as reflections of basic social forces and social structures. The personal following of the Kennedy family in Massachusetts made a discernable difference. But the supposed independence of Ohio voters, the strong backing of Kennedy by the UAW in Michigan and by the party organization in Connecticut, or the power of the machine in its crudest form in Chicago, did not make these states vote very differently from the way they would have voted anyway, given their population composition.[23]

---

attempted to treat as Western. The assumption that the Mountain area was uniform with the West was misleading.

[23] The acute reader might note that political party was one of the cell-defining variables along with sex, SES, rural-urban residence, and so on. If party had been used in making synthetic states we would have been smuggling political history back into our simulations by a back door, for we would have been using an estimate for each state of the number of Republicans, Democrats, and Independents in it. We did not do that, however. We had no source for estimating party affiliation state by state. The census, our main source of state data, does not tell us party sizes. Use of election returns, by averaging over many contests, was a possibility, but since some of the issue-clusters

SIMULATING THE 1960 CAMPAIGN

The Problem Posed

In August 1960, when we were called upon to use the system that had been assembled, the most pressing question facing the Democratic high command was how to handle the religious issue. Kennedy had handled it in Wisconsin and West Virginia forthrightly by public attacks on bigotry. In both cases he was successful. He had gone on to win the nomination in a splurge of optimism bolstered in June by straw polls showing him in the lead. Then suddenly right after the convention the bottom seemed to fall out. The polls began showing Nixon well ahead and everywhere anti-Catholic activities, statements, and literature began to appear.

The number one question at the Democratic National Committee, in those adverse circumstances as in any campaign, was who would win. But from the start of the Simulmatics project we had insisted that we would not address ourselves to that question. We did not ourselves know how good an absolute vote predictor our new approach would prove to be. More important than that, the question "who will win" is an empty question. The question embodies all the anxieties and tensions of a campaign. The candidate cannot bear the uncertainty of waiting for election day. He

---

used in the simulations were to include past votes, this seemed a dubious procedure. So what we did was to assume that *within a voter-type,* party affiliation was distributed in the same proportions for all states in a region. The regional party affiliation breakdown which our data bank had for each voter-type was applied to that voter-type in creating each synthetic state. That means that the number of voter-types whose frequency in each state we sought to ascertain from independent sources such as the census was really 36 maximum, not 108. A synthetic state was really a distribution of these 36 types.

wants the poller to relieve his uncertainties by providing a prediction. A reassuring one may be psychotherapeutic for him. But aside from that psychic benefit, the information is useless. No action consequences follow from either a favorable or unfavorable reply. One must go on with the fight in either case. So we had ruled out any attempt to predict the answer to the number one question and had defined our role as estimating the relative gain or loss to be obtained from adopting one strategic alternative or another.

The strategic alternative of primary interest was between a policy of minimizing attention to the religious issue by discussing it as little as possible and a policy such as that already adopted by Kennedy in West Virginia of publicly attacking those who introduced the issue of the religion of a candidate into the campaign. The latter alternative, however much it might strengthen opposition to bigotry, also tended to draw Kennedy's religion into the limelight. Every attack on those who injected the religious issue in fact brought it even more into focus.

Both Kennedy and Nixon apparently felt that that strategy which focused on the issue by explicitly asserting its irrelevance helped the Democrats. Nixon protested that Kennedy by attacking bigotry was focusing attention on a question which had no place in the campaign. Nixon proposed a ground rule that neither candidate would even mention the existence of the issue. He ordered all Republican organizations to act in that way. They were not to permit any reference to Kennedy's Catholicism in any context, whether by way of appeal to prejudice or by way of rebuttal of prejudiced statements. One Republican official told us about his embarrassment at not being able to silence a bigoted local leader of his own party because his instructions were so strict that he could not mention the issue even in reply.

While Kennedy and his close advisors were inclined to the strategy of meeting prejudice overtly, they were far from sure that they were right. The policy question in August 1960 was what would happen if the religious issue became sharply acerbated. What would happen if it became the dominant issue of the campaign, which it easily could if prejudice continued to rise, as indeed it seemed to be doing, and if the candidate also responded in the open forum? It was to this question that we addressed ourselves.

The Formulas Chosen

When faced with a question like that about a hypothetical future, the next step in a simulation is to formulate subjectively, but with as much detail and precision as possible, what one believes will be true under the alternative being simulated. One may draw on theory, experience, common sense, guesswork, or any other source. One must, however, formulate quite rigorously the principles that one expects will become operative. The simulation does not provide these insights. It works out their implications in a large system.

The notions we had about the impact of the religious issue on people of different voter-types were expressed in a series of equations that we will now review. These equations were designed to represent a campaign in which only two forces played any role: one, past voting inclinations, and the other, attitudes toward a Catholic in the White House. Data about each of these two factors were included in some issue-clusters in our data bank. There were various past vote-clusters, such as one on how people said they had voted in 1956, one on how they said they had voted in Congressional elections; and there was a cluster on how they felt about having a Catholic in the White House. Thus for each of the 480 voter-types we had an objective empirical measure of its position on some relevant issue-clusters.

The basic mechanism in the simulation was a fairly straight-forward application of the cross-pressure findings of the earlier election studies. We grouped our set of 480 voter-types into 9 possible cross-pressure subsets arising from a 3×3 classification on religion and party: Protestants, Catholics, and Other; Republicans, Democrats, and Independents.

Into each of these nine situations fell a certain number of the 480 voter-types. The two main groups containing people under cross pressure on the religious issue were those marked with an X in Table 1.3, the Protestant Democrats and the Catholic Republicans.

TABLE 1.3
CROSS-PRESSURE PATTERNS

	Republicans	Democrats	Independents
Protestants	(1)	X (2)	(2)
Catholics	X (4)	(3)	(3)
Other	(4)	(5)	(5)

For each of the nine situations we made a prediction about how voters in that situation would behave. Actually some of the situations seemed similar so we ended up with only five separate sets of predictive equations. We shall go through these in order. (The relevant equations are indicated by the numbers written in the boxes of the ninefold Table 1.3.)

1. *Protestant Republicans.* They were not under cross pressure. Since our data bank revealed no substantial dislike of Nixon as an individual among such voters, we did not expect their vote in 1960 to differ substantially from their vote in 1956, even though Eisenhower was not running. Thus for them we wrote two equations

$$V_k = P_{56}(1 - P_{35})$$
$$V_n = Q_{56}(1 - P_{35})$$

meaning that the predicted Kennedy percentage (V_k) in any voter-type of this Protestant-Republican sort would be the percentage of persons in that voter-type who indicated a preference for Stevenson in 1956 (P_{56}) reduced by the nonvoting record of that voter-type $(1 - P_{35})$. The equation for the expected Nixon percentage (V_n) was the same except that it used the 1956 Eisenhower supporters (Q_{56}).

In short we said that Protestant Republicans would vote exactly as they did last time, mostly voting Republican, a few odd ones—just as many as last time—voting Democratic. But before adding the votes arising from those voter-types to the votes of other voter-types we deflated a little the claimed vote on both sides to allow for the fact that turnout in a real election is always less than "voting" as it appears on a straw poll.[24]

2. *Protestant Democrats and Protestant Independents.* The number 1 set of equations was the simplest set used. Let us now turn to a more complicated set, that for a group under cross pressure—Protestant Democrats. First, we decided that, barring the religious issue, Congressional vote intentions would be a better

[24] William A. Glaser, "Intention and Voter Turnout," *American Political Science Review*, Vol. 52 (1958), pp. 1030–1040.

index of the Protestant Democrats' 1960 vote than would their 1956 vote intentions. Too many of them were Eisenhower defectors in 1956 for us to believe that 1956 was a good indicator of normal behavior for them. We believed that a better predictor of the 1960 vote of a Protestant Democrat would be how he had voted for Congress in the past modified by any anti-Catholic sentiment he might have. To say the same thing in another way, we believed that the relatively firm party-line vote which occurs in Congressional years would be a better base on which to make a correction for the religious issue than the 1956 vote when Eisenhower's glamor attracted many Democrats.

The most obvious equation to have used for Protestant Democrats was one which subtracted from the Democratic Congressional vote a proportionate part of those voters who on the anti-Catholicism issue-cluster expressed anti-Catholic sentiments.[25] However, this easy equation might give a false result, for it might be that those very Democrats who were anti-Catholic were the ones who in practice voted Republican anyway. In short, a question was: Were the bigot defectors among nominal Democrats mainly right-wingers whose votes the Democrats would lose even without a Catholic candidate? Our system could not give us that information for each respondent incorporated into our data.[26] While one respondent in a voter-type might have been polled in a survey in 1958 about his vote intentions, another man of the same voter-type, on a different survey, might have been polled on whether he would vote

[25] The typical question that went into this cluster was "If your party nominated an otherwise well-qualified man for President, but he happened to be a Catholic, would you vote for him?"

[26] This point, which is important but technical, is developed more fully in the Appendix to this chapter.

for a Catholic for President. To estimate the correlation between these two variables we had to find one or more surveys on which both questions appeared. We then ran anti-Catholicism by Congressional vote for each of the more numerous Protestant Democrat voter-types. We found that among them the ratio ad/bc in the fourfold Table 1.4 averaged about .6. With that in-

TABLE 1.4

CORRELATION TABLE FOR ESTIMATING IMPACT OF RELIGIOUS ISSUE ON A DEMOCRATIC VOTER-TYPE

Congressional Vote Intentions	Attitude on the Religious Issue	
	Anti-Catholic	Not Anti-Catholic
Democratic	a	b
Republican	c	d

formation we could estimate how many of the anti-Catholics were hopeless cases anyhow (that is, had gone Republican for Congress) and how many would be net losses only in a campaign dominated by the religious issue.

It should be added here that we decided to take poll replies on the religious issue at face value. We were not so naive as to believe that this was realistic, but since we were not trying to predict absolute voter percentages, but only relative ones, all that mattered was that the true extent of anti-Catholicism voter-type by voter-type should be linearly related to the percentage overtly expressed. Even this could be assumed as only a promising guess.

Finally, in predicting the vote of the Protestant Democrat voter-types, we took account of the previously established finding that voters under cross pressure stay home on election day more often than voters whose pressures are consistent. Therefore, for our 1960 estimate we doubled the historically established nonvoting index for these types.

Thus we arrived at equations applied to each Protestant Democrat voter-type:

$$V_k = (P_{58} - a)(1 - 2P_{35})$$
$$V_n = (Q_{58} + a)(1 - 2P_{35})$$

The estimate of anti-Catholics among Democratic Congressional voters (that is, persons in cell a in the fourfold Table 1.4) was arrived at by the computer, given

$a + b = P_{58}$ = per cent of the voter-type who had indicated a Democratic Congressional vote intention.

$c + d = Q_{58}$ = per cent of the voter-type who had indicated a Republican Congressional vote intention.

$$a + c = P_{14}(P_{58} + Q_{58})$$

P_{14} = per·cent anti-Catholic.

$$\frac{ad}{bc} = .6$$

Independents cannot be said to be under cross pressure if they are true Independents. But few Independents are truly independent. They have propensities one way or the other which vary with their social milieu. Among those with a propensity to vote Democratic this inclination may be offset by anti-Catholicism.

The only respect in which one might doubt the applicability of the equation for Independents was the large nonvoting factor applied to party Democrats because of the extreme cross pressure upon them. But Independents are less interested in politics and vote less anyhow, so even on that point the equations could be justified.

3. *Catholic Democrats and Independents.* These are again groups who would feel no cross pressure. Their basic vote propensity for 1960 we again expected to be indexed better by their Congressional vote than by their Eisenhower-Stevenson vote in 1956. However, if the religious issue were maximally salient (and that is what we were simulating) we expected that they would be even more solidly Democratic than they were when voting for Representatives. We had to guess in the absence of better evidence. We had no issue-cluster dealing with pro-Catholicism by Catholics, only one dealing with anti-Catholicism. So our guess was intuitive at best. We guessed that one third of the vote among Catholic Democrats and Independents which the Democrats normally lost in Congressional contests would come back to them in a Kennedy campaign where the religious issue was at the top of attention.

This is the relationship expressed by the first term of the next two equations—one third of the defectors in the Congressional election come back. The second term is the usual turnout correction.

$$V_k = \left(P_{58} + \frac{Q_{58}}{3} \right) (1 - P_{35})$$

$$V_n = \frac{2Q_{58}}{3} (1 - P_{35})$$

4. *Catholic and Other Republicans.* Catholic Republicans would be under cross pressure in a campaign where a Catholic Democrat was under attack for his religion. So in an indirect way would be a Jewish or Negro Republican since he would sense bigotry as an attack on him.

At first glance the next equation seems to have these cross-pressured Catholic Republicans acting like the Catholic Democrats because again, as a guess, we postulated that one third of those who voted for Republican Congressional candidates would swing over to the Democrats in the simulated environment. But note that the one-third shift for type 3 above is among a small group: Democrats who despite their party had favored a Republican for Congress. Here the one-third shift is among a large group: Republicans who voted Republican. It will be recalled that in stating the hypotheses of cross-pressure theory we asserted that most voters resolve their cross pressures by staying with their party rather than switching. A one-third shift is a big shift.

Another hypothesis said cross-pressured voters vote less. They stay home. In the light of indirect evidence to be presented later, it is not fully clear that this is true, but we initially assumed it. We furthermore assumed that those who considered themselves for Nixon would stay home more than those who considered themselves for Kennedy since to the former their vote would seem almost an act of alliance with the bigots against themselves. We raised the nonvoting factor by 2 and 3, respectively.

$$V_k = \left(P_{58} + \frac{Q_{58}}{3} \right) (1 - 2P_{35})$$

$$V_n = \frac{2Q_{58}}{3} (1 - 3P_{35})$$

5. *Negro and Jewish Democrats and Independents.* The remaining groups are not large. Table 1.5 shows the relative number of voters in each condition we are distinguishing. But while the numbers of Negro and Jewish Democrats and Independents were not large, formulating the appropriate equations for these groups was highly problematic.[27]

---

[27] Strictly, what we are here calling "Negro and Jewish" should be labeled "Other." It is a residual category containing all persons not

We decided to treat them as under cross pressure. We had already done that with Negro and Jewish Republicans. We had postulated that while most of those Republicans would continue to vote Republican some would resent the bigotry of the campaign we were simulating and would switch. Among Negro and Jewish

TABLE 1.5
NUMBERS OF POTENTIAL VOTERS IN DIFFERENT
CROSS-PRESSURE SITUATIONS

	Republicans	Democrats	Independents	Total
Protestant	22.9%	25.2%	13.9%	62.0%
Catholic	5.0%	12.3%	6.0%	23.3%
Other	3.0%	7.5%	4.2%	14.7%
Total	30.9%	45.0%	24.1%	100.0%

Republicans we could disregard those who were anti-Catholic for they were being pressed to their normal Republican vote anyhow, but those who were anti-anti-Catholic would be under cross pressure. Now for Negro and Jewish Democrats the situation was reversed. Those who resented anti-Catholicism would be pressed in the direction they would follow anyhow and could be disregarded. Those who were themselves anti-Catholic would, however, be under cross pressure. So we introduced into the equations for them the same sort of cross-pressure mechanism that we used for Protestant Democrats.

For Negro and Jewish Independents the situation was even fuzzier. While it was clear that some part of these voter-types should be treated as under some cross pressure between normally Democratic vote in-

---

classified in our system as Protestants or Catholics. But to all intents this means Negroes (for whom we disregarded religion) and Jews. Other-religion and no-religion voter-types are negligible.

clinations (despite their independence) and the religious biases of some of them, it was not clear what election to use as a base in assessing the degree of their Democratic vote inclinations. (The anti-Catholicism issue-cluster in the data bank gave an acceptable measure of the degree of anti-Catholicism of each votertype.) Jewish and Negro Democrats were not as much swept out of their normal patterns by the Eisenhower appeal in 1956 as were the Protestants and especially the Catholics. It could be argued that given the lesser Ike distortion in 1956 among Jews and Negroes, using Presidential year voting as a predictor for another Presidential year was better than using Congressional voting. We hedged by averaging between 1956 and Congressional figures.

Finally, we estimated turnout for these groups to be poorer than par, since Independents never turn out well and Negroes are extensively disenfranchised.

$$\frac{ad}{bc} = .6$$

$$a + b = \frac{P_{58} + P_{56}}{2}$$

$$c + d = \frac{Q_{58} + Q_{56}}{2}$$

$$a + c = P_{14}\left(\frac{P_{58} + Q_{58} + P_{56} + Q_{56}}{2}\right)$$

$$b + d = (1 - P_{14})\left(\frac{P_{58} + Q_{58} + P_{56} + Q_{56}}{2}\right)$$

$$V_k = \left(\frac{P_{58} + P_{56}}{2} - a\right)(1 - 2P_{35})$$

$$V_n = \left(\frac{Q_{58} + Q_{56}}{2} + a\right)(1 - 2P_{35})$$

Using common sense, social science theory, and similar guides, we thus expressed our best judgment

TABLE 1.6

EQUATIONS USED IN AUGUST 1960 SIMULATION

Protestant Republicans:	$V_k = P_{56}(1 - P_{35})$
	$V_n = Q_{56}(1 - P_{35})$

Protestant Democrats;
Protestant Independents:

$$\frac{ad}{bc} = .6$$

$$a + b = P_{58}$$
$$c + d = Q_{58}$$
$$a + c = P_{14}(P_{58} + Q_{58})$$
$$V_k = (P_{58} - a)(1 - 2P_{35})$$
$$V_n = (Q_{58} + a)(1 - 2P_{35})$$

Catholic Democrats;
Catholic Independents:

$$V_k = \left(P_{58} + \frac{Q_{58}}{3} \right)(1 - P_{35})$$

$$V_n = \frac{2Q_{58}}{3}(1 - P_{35})$$

Catholic Republicans;
Other Republicans:

$$V_k = \left(P_{58} + \frac{Q_{58}}{3} \right)(1 - 2P_{35})$$

$$V_n = \frac{2Q_{58}}{3}(1 - 3P_{35})$$

All Others:

$$\frac{ad}{bc} = .6$$

$$a + b = \frac{P_{58} + P_{56}}{2}$$

$$c + d = \frac{Q_{58} + Q_{56}}{2}$$

$$a + c = P_{14} \left(\frac{P_{58} + Q_{58} + P_{56} + Q_{56}}{2} \right)$$

$$b + d = (1 - P_{14}) \left(\frac{P_{58} + Q_{58} + P_{56} + Q_{56}}{2} \right)$$

$$V_k = \left(\frac{P_{58} + P_{56}}{2} - a \right)(1 - 2P_{35})$$

$$V_n = \left(\frac{Q_{58} + Q_{56}}{2} + a \right)(1 - 2P_{35})$$

Note: Definition of symbols

P_{14} = per cent anti-Catholic.

P_{58} = per cent of the voter-type who indicated casting a Democratic Congressional vote.

Q_{58} = per cent of the voter-type who indicated casting a Republican Congressional vote.

P_{56} = per cent of the voter-type who indicated casting a 1956 Stevenson vote.

Q_{56} = per cent of the voter-type who indicated casting a 1956 Eisenhower vote.

P_{35} = per cent of the voter-type who indicated being nonvoter.

on how different types of voters would respond to a campaign focused on the religious issue. But having codified our best judgment was not the same thing as estimating the consequences of these judgments. The computer applied these equations, which represented our judgments, to the parameter values on past vote, nonvoting, and religious prejudice recorded in the data bank for each of the 480 voter-types; it then computed weighted averages for each of 48 states. That gave us our simulation result.

In Table 1.6 we recapitulate all the relevant equations.

The Outcome of the Simulation

The report that we made to the Democratic National Committee on our simulation of the consequences of focusing on the religious issue was limited to the North because the assumptions made about party voting would not apply to some states in the South where voters' decisions to vote for the Democratic electors or the Republican depend on the relation of the Democratic organization to the national ticket. A Democratic vote would not necessarily be a Kennedy vote everywhere in the South. The outcome was a ranking of 32 states ranging from the one in which we estimated Kennedy would do best to the one in which we estimated he would do worst. The ranking was:

1. Rhode Island	8. California	15. Minnesota
2. Massachusetts	9. Arizona	16. Missouri
3. New Mexico	10. Michigan	17. Pennsylvania
4. Connecticut	11. Wisconsin	18. Nevada
5. New York	12. Colorado	19. Washington
6. Illinois	13. Ohio	20. New Hampshire
7. New Jersey	14. Montana	21. Wyoming

22. Oregon	26. South Dakota	30. Utah
23. North Dakota	27. Vermont	31. Idaho
24. Nebraska	28. Iowa	32. Maine
25. Indiana	29. Kansas	

The product-moment correlation over states between the Kennedy index on the simulation (not strictly speaking a per cent of the vote) and the actual Kennedy vote in the election was .82. It should be emphasized that this satisfying result was based upon political data not a single item of which was later than October 1958. Surveys on the 1960 election were not available soon enough to be incorporated into this analysis.

How good a result is that? We shall see in the next chapter that equally good predictions might have been possible by other means, but what of the obvious means, straw-vote results? The most relevant comparison is with the Kennedy-Nixon trial heats on polls taken at the same time as the latest polls used in the simulation. The correlation between the state-by-state result of these polls and the actual outcome is but .53 as compared to .82 for the simulation. The simulation, in short, portrayed trends that actually took place between the time the data were collected and election day two years later. By election day the polls were doing very well indeed. Their final national predictions were very close to the national result. State-by-state reconstructions, presuming such to have been available, would also have been excellent at that time. But one had to simulate certain processes of change in order to get good predictions out of the earlier polls.

At some point in the history of the campaign, poll data came into close correlation with the November election results and thus with our simulation too. The date when the raw poll results became as or more predictive than the simulation would be the date when the mechanisms of voter behavior anticipated in the simulation had become reality.

A simulation is not as good as a poll if the problem is to learn how voters have already made up their minds. The way to learn that fact is to ask it. But the simulation did take old data collected before the voters had made up their minds and acted out how they would make up their minds before they did so.

This does not mean that what we did in 1960 we could equally well have done in 1958. Although our data were all there to be used in 1958, one essential fact was missing at that time, namely a sensible judgment about what scenario to simulate. It was not clear until 1959 or 1960 what sort of campaign would occur in 1960. The predispositions of the public were all there and discoverable two years earlier. Buried in the public opinion poll records of 1952–1958 were data that would permit the estimation of voter behavior in 1960 for most particular campaigns that might have occurred. But there was no way of knowing what campaign might occur. To anticipate which campaign to simulate required that one be nearer the events. In 1958 a prediction of the candidates might have included Stevenson, or Johnson, or Rockefeller. The issue might not have been the religious issue. The campaign might have occurred in a depression, or national attention might have been absorbed by a new war somewhere in the world. In short, many scenarios were possible for 1960 and most were simulable. It was only when the events were close at hand, however, that a realistic scenario could be chosen for the simulation runs.

When we reported the results of our August 1960 runs, we reported them in a rank order form. We listed the 32 Northern states in the order of where Kennedy should do best in the circumstances simulated to where he should do worst. We refused to treat our index number as a per cent of the vote. In fact it was a per cent. Each of the variables that entered our calculation was a per cent: the per cent of the voters in our data bank who in 1956 were reported for Stevenson and for Eisenhower, who were reported for Democratic or Republican Congressional candidates, who were reported as not registered to vote, and who were recorded as not wanting a Catholic for President. But caution made us wonder whether the arbitrary weights we were using on nonvoting, on the shift of Catholics, and on the honesty of reporting anti-Catholicism would permit a prediction of absolute vote values.

In retrospect we should have tried to be more daring. Our system did reasonably well in predicting absolute percentages and might have done better had we sought to design it to estimate actual vote outcomes. The square root of the mean squared deviation, state by state, using the 4-year master was 5.4 per cent with a median absolute difference of 4.4 per cent. That is only a fair result. It will be recalled that originally we had designed our equations to be used on a 5-year master with data through 1959. That was not usable in the two weeks we had available for reporting and so it was not used. But with that added data which would have been more up to date we would have predicted with a root-mean-square error of 3.9 per cent and a median absolute error of 3.25 per cent state by state. The reasons for this improvement are discussed in the next chapter. Reasonably close prediction seems possible. Now with hindsight we can explore as guidance for future attempts how we might have done better. That is the problem to which we now turn.

APPENDIX TO CHAPTER 1

There is a more serious methodological difficulty in using multiple-survey sources than that of the comparability of the different question wordings used on different surveys, namely the problem of estimating correlations between variables that have been explored in separate studies. Suppose that two variables A and B are measured on separate surveys, but each correlates with a third variable C which is measured on both, then A and B will be correlated if the measure is their distribution across C, but across individuals they may be uncorrelated or differently correlated. A general rule is that correlations across respondent-types can be measured regardless of the number of surveys involved, while correlations within respondent-types cannot be measured unless both variables happen to occur on a single survey.

For example, illiteracy is negatively correlated with socioeconomic status since illiterates are found at the bottom of the ladder. Reading of lowbrow magazines (for example, movie magazines in the United States) is also negatively correlated with socioeconomic status. So across socioeconomic-status levels illiteracy and reading of movie magazines are correlated. But across individuals they are not. It is not illiterates who read movie magazines.

The example is instructive. A crude breakdown of socioeconomic-status levels, say a mere dichotomy, would show illiteracy and movie magazine reading to be highly correlated for both would appear in the bottom half of the population. The finer the breakdown, the more likely that this correlation (which does not exist for individuals) will disappear in the across-type correlation too. Note that if we break socioeconomic-status levels finely enough, or even better if we define our population types by both socioeconomic-status and education, we will soon discover that illiteracy occurs at the very very bottom of the socioeconomic-status distribution and among people with virtually no education while movie magazine reading occurs most among people of moderately low socioeconomic-status and those with six or more years of education.

Nonetheless, the fact remains that a correlation across types never proves a correlation across individuals. One can improve the chances that such an inference is valid by making the types finer and including more variables in the definition of them. The variables most often used to define respondent types are the basic demographic items which are reported on the face sheet. Standard face-sheet items—such as sex, occupation, age, place of residence—occur on almost all surveys and the reason they do is because they are powerful determinants of many kinds of behavior.

As a general rule, if one controls enough of the face-sheet variables, there is a very good chance that inter-variable correlations within respondent-types will disappear.

So we may expect within-type correlations almost always to be very much smaller than those that occur in the population as a whole because correlations among social indices are largely accounted for by joint correlations of the indices with other social variables.

But it is not necessarily the case that within-type correlations will disappear. Intelligence, for example, correlates with many forms of behavior even when one holds education constant. Admiration for the Soviet Union and dislike of private enterprise will continue to correlate regardless of how many socioeconomic and demographic variables one holds constant. In short, one cannot simply assume away intratype correlations among variables, even though they will usually be small if enough other factors are held constant.

In the Simulmatics project, interissue correlations were a serious problem. We met the problem in part by having as many as 480 types, each defined by a number of major variables. Thus, the types were socially quite homogeneous, and many demographic variables that could introduce correlations were being controlled for. In addition, there was a way, when it seemed important, roughly to estimate within-type correlations for most pairs of variables.

Since we were working with a large collection of public opinion poll data and since each issue was indexed by questions from a number of different polls, there was a fairly high probability that we could find some one survey at least, on which at least one question had been asked which was a component of each of the two indices which we wished to correlate. Whenever that turned out to be the case the problem was solvable. Having found a single survey with questions on both issues, then we could run correlations within voting types.

Needless to say these correlations were individually worthless since with 480 voter-types and with only a couple of thousand interviews in a single survey the number of cases for any one correlation would be trivial. There were two options. One was to aggregate voter-types on any characteristic which seemed irrelevant to or uncorrelated with the particular variables being studied. The other was to disregard the significance of the individual correlations. All we were interested in was an estimate of their average.

With some appropriate compromise between these two options it becomes possible to arrive at a rough estimate of the typical correlation within the voter-types. We could assign values to the cells of the contingency table whose marginals we already knew and whose correlation coefficient we now have estimated.

If two variables are uncorrelated across individuals the expected values for the cell entries in a matrix relating two variables is the cross product; for example, $a = [(a + b)(a + c)]/(a + b + c + d)$. However, in the presence of correlation that is not so. Some correlation measure such as the expression ad/bc has to be used along with the marginals to solve for each cell.

2

Testing the Assumptions

INTRODUCTION

A complex model can predict real-world outcomes correctly and yet be wrong in many details. It may predict accurately because the main effects are correctly represented and yet the model may contain many irrelevancies. So one always must question the details of a complex model, even if it passes the test of good prediction.

The simulation we used in the 1960 campaign did, as we have seen, prognosticate the vote reasonably well. Still it may legitimately be asked what in the equations accounted for this success, and whether there were parts of the equations that contributed nothing or even did harm. It is to this problem that we now address ourselves.

The exploration of the relative importance of different parts of the model is technically called "sensitivity testing." Sensitivity testing is one of the more important uses of computer simulation. One often designs a computer model not for the purpose of predicting but for the purpose of gaining an understanding of the process represented. By varying each of the parameters of the model one can see which ones make a difference and which ones do not. Some few variables may account for much of the variance, some may account for little. To explore the sensitivity of the prediction to the value of each parameter is one way of gaining a deeper understanding of what is taking place. We shall now examine the variables that went into our 1960 equations and see what would have happened if each had been a little bit different.

CHOICE OF DATA BASE

The results obtained in a simulation are, of course, sensitive to the time and place from which the data come. For this study, the place of origin of the data was no problem because all the 65 polls with which we worked were national sample surveys. But time was a problem. Our data spanned the years from 1950 through 1959. Many people have questioned whether we should have used 1952 and 1954 election survey data at all. Query number one then is whether our results were sensitive to the dates of the data included. Would we have done better using only the last or the last two of the five biennial periods into which the data bank was divided?

The 5-period cumulative data bank, containing all 65 polls, is substantially more accurate in predicting the election than any other data base for the simulation which could be formed from any subset of the periods. In fact the simulation results obtained from using the latest surveys only—those from the fifth period—are extremely inaccurate due to the small amount of information they contained on anti-Catholicism and the overstatement in them of true Democratic strength on Congressional voting, arising from the landslide character of the 1958 election.

Data on four issues were used in the simulation equations: the 1956 Presidential vote, Congressional voting, anti-Catholicism, and turnout. No single one of the five biennial matrices has sufficient information on anti-Catholicism. It was necessary to use all available data on this issue since the polls seldom asked about it before the 1958 contest. However, each of the five biennial matrices had extensive information on Congressional voting, and all but one of the biennial matrices, 1951–1952, had information on voting in the 1956 Presidential election. None of the various plausible alternatives such as using the 4- or 5-period cumulative matrix for Presidential vote, anti-Catholicism, and turnout, but using the 1957–1958 or 1959 matrix for Congressional vote,[1] proved to be as good as simply using the 5-period master, the procedure that had been called

[1] The more complex alternative which uses the 1958 Congressional vote instead of the 4- or 5-election average correlates with the November results only .7 instead of .8 which is obtained with either the 4- or 5-year masters. It should be noted, however, that the 1958 Congressional vote substituted for Congressional voting across the decade, pulls up the Democratic percentage since 1958 was a Democratic year. It thus produces a smaller mean error since our simulation underestimated Kennedy's vote. However, as already noted, we did not try to make our simulation correspond to correct absolute levels. There is a real dilemma in the two alternative criteria of the concept of good fit. Our point of view is that an adequate correlation of simulation results with actual results is the first test that a simulation must pass, for only if there is a good correlation is it plausible that the mechanisms in the model correspond with those in the real world. As between models which correlate adequately well with the real results—but only between them—absolute deviation becomes a relevant criterion, and a more rigorous one.

Note that throughout this section, unless otherwise noted, we are discussing only the 32 states of the North, which are what we used in our original simulation.

for in the original simulation plan. The benefits of large numbers and broadly based coverage proved to be greater than the benefits of timeliness. One-hundred-thirty-thousand interviews distributed over time proved to provide a better base than would one randomly timed poll with 130,000 cases. And 65 polls from a decade proved better than the last 20 of them.

ANTI-CATHOLICISM

Intuitively the most interesting variable in our simulation was anti-Catholicism. Indeed, the simulation was designed to explore its effects. But the mere fact that the simulation gave a good prediction does not prove that anti-Catholicism was an important factor in determining the election outcome. In simplest terms what our simulation did was to take a normal vote base—basically the Congressional vote—and modify it by religious factors. If perchance the Congressional vote base correlated with the 1960 outcome and religious bias had a uniform effect in every state (which could but need not be a zero effect), then we would get a good prediction from the simulation despite the triviality of the effect of the religious issue.

Was that the case? To find out we must see how well the simulation would have predicted without the inclusion of the religious factor. Was the level of correlation between the simulation and the real outcome as high for a simulation that did not include the religious factor as for one that did include it? The answer is a decided "no." The outcome of the simulation is highly sensitive to the factor of religion.

We document this conclusion first by seeing what happens if we disregard anti-Catholicism. Elimination of the anti-Catholicism issue-cluster from our equations would have lowered the correlation with the November results from .82 to .62 which is only about 60 per cent as good.[2]

CATHOLIC SHIFT

The religious issue cut both ways. Not only did some Protestants reject Kennedy but also some Catholic Republicans swung to him. We have seen that the net result was highly sensitive to the Protestant bias factor. Now let us see how sensitive it was to Catholic shifts.

The original simulation assumed that one third of all Catholics who would otherwise have voted Republican would swing to Kennedy. That was an arbitrary figure arrived at by guess. In the data bank we had no set of empirical values arising from polls to tell us how strong was the pro-Catholic bias of each Catholic voter-type. We had no figures comparable with the set of figures we had on anti-Catholic bias of each Protestant voter-type. The pollers had asked Protestants whether they would object to voting for a Catholic for President. They had not asked Catholics whether they would vote for a Catholic candidate just because he was a Catholic (and for good reasons). So while we had a plausible shift expectation for each of the 285 non-Catholic voter-types, we had no basis but guess, in the form of an across-the-board assumed shift, for all 195 Catholic voter-types. The value that we set for the Catholic shift was thus arbitrary. It is, therefore, legitimate to enquire whether our guess about its value was a reasonable one. Instead of testing what would have happened if we had dropped the factor of a Catholic shift altogether, we test what would

[2] The variance accounted for is the square of the correlation coefficient, that is, about .66 and .38, respectively.

have happened if we set it at each value from 0.1 to 1.0. In short, we ask not only what would have happened if .333 of Catholic otherwise-Republican voters shifted, but also what would have happened if one tenth did, two tenths did, more, or all did. Figure 2.1 shows

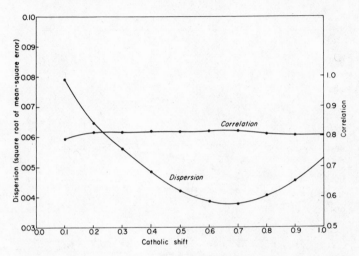

FIGURE 2.1 How good would the simulation have been with different values of the Catholic shift (using the 4-period master)?

the result of varying the Catholic shift on the 4-period data matrix used in the original simulation. The correlation coefficient is quite stable for all values of the "C-shift," but the dispersion (the square root of the mean-square error) varies markedly, achieving its lowest level, .038 per cent, when the "C-shift" is .7, in other words when it is assumed that 70 per cent of Catholic Republicans shifted to Kennedy. This figure seems much too high intuitively, and in the next section we discuss a modification of the simulation producing a more reasonable "C-shift" to correspond with the best over-all data fit.

Figure 2.2 carries the analysis of the sensitivity of the results to the religious issue parameters one step further. It illustrates what happens when we simultaneously vary the "C-shift" and anti-Catholicism. How good a correlation do we get, for example, if we drop

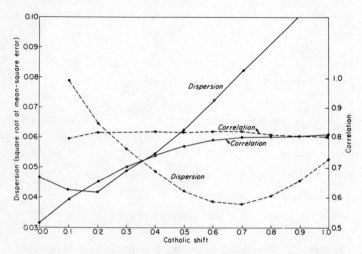

FIGURE 2.2 The effect of anti-Catholicism and the Catholic shift simultaneously compared (using the 4-period master).

‒ ‒ ‒ ‒ ‒ ‒ = Using anti-Catholicism issue-cluster

——————— = Omitting anti-Catholicism issue-cluster

them both and simply use the party base in the equations as a prediction of the 1960 vote? The result is not very good. The correlation is about .5 or less than 40 per cent as good as the original simulation.

On the other hand we can assume no anti-Catholicism among Protestants and still get a good correlation with the actual results but only if we assume that between 80 per cent and 100 per cent of all Catholics who would otherwise have voted Republican voted for Kennedy. And that pair of assumptions produces an average deviation from the real results of between 8 and 12 percentage points. It separates states with a large Catholic population from those with a small, well enough to produce a close correlation of the predicted to the actual Kennedy vote,[3] but it does it in a way that does violence to the real-world process. It does it in a way that denies to Nixon many of the votes he got.

Clearly there is no simulation that omits anti-Catholicism which is better than the original simulation.[4] Nor is there any that gives a good prediction which fails to recognize that a large Catholic shift also took place.

The Catholic shift worked in the same way on the 5-period data bank. Figure 2.3 compares the 4- and 5-period simulations at all values of the C-shift. While the correlations are always essentially the same, the

[3] The state-by-state correlation between the actual Kennedy vote outside the South and the Catholic population in the state was .68.

[4] One might legitimately ask what effect varying the magnitude of the anti-Catholic shift might have had. We might have written into the equation that the Protestant shift would be .8, or 1.2 or 1.3 times the value found in the data bank for the anti-Catholicism issue-cluster, instead of 1.0 times its value which is the value we used. To examine this, however, would be redundant. It is clear that as far as the correlation is concerned, we are doing the same thing by varying the C-shift. Whether we pull the Protestant norm away from the Catholic norm or the Catholic norm away from the Protestant norm we are doing virtually the same thing. We are separating states with large Catholic population from states with small ones and thereby improving the correlation. Similarly it is clear that raising the Protestant shift above 1.0 while improving the correlation (at least at low values of the C-shift) would make the dispersion worse by increasing Nixon's predicted vote, which was already predicted too high. In short, at factors of anti-Catholicism above 1.0 times the data bank figure, the left ends of both the correlation and dispersion curves would be higher on the vertical axis. On the other hand lowering the factor below 1.0 times the data bank figure would simply bring the dotted curve nearer to the solid one. The correlation curve would be near the solid line curve for a longer range and the nadir of the dispersion curve would be higher.

5-period simulation has a much lower square root of mean-square error for any Catholic shift below .6.

The 5-period consolidated data bank contained more poll replies on the 1958 election which the Democrats won. It therefore gave them a better showing, less influenced by the Eisenhower bandwagon of 1952 and

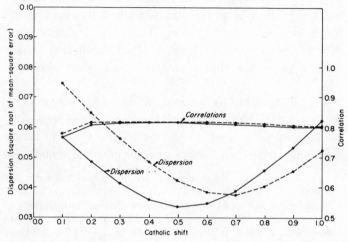

FIGURE 2.3 Comparison of results with 4-period and 5-period master matrix.

$-----$ = 4-period master matrix
$————$ = 5-period master matrix

1956. For those reasons it produced a more reasonable result. A good prediction is achieved with a Catholic shift of only .5 of the Catholics who would otherwise vote Republican—a much more plausible value than the .7 or .8 or .9 or 1.0 which we found to be good predictors just above.

PARTICIPATION PARAMETERS

Varying the participation and turnout parameters on the original simulation suggests that the turnout pattern for the 1960 election was substantially different from that in 1952 and 1956. More specifically, it seems that the Democratic turnout rate was much higher than usual, in fact probably as high as for Republicans. The equations in the original simulation all involved a term that was either $(1 - P_{35})$, $(1 - 2P_{35})$, or $(1 - 3P_{35})$, where P_{35} is the percentage of the voter-type who said they normally did not vote or register and the weighting factor, that is, 1, 2, or 3, was determined by the relative amount of cross pressure the voter-type was assumed to be under.

Instead of estimating turnout for the voter-types by the issue-cluster that reported normal nonregistration, we could also estimate turnout for each type by the relative saliency of the election for them as evidenced by the issue-cluster dealing with interest in the election. Or, we could estimate turnout by both factors, nonregistration and interest in the campaign, or we could use neither factor, that is, assume turnout equal for all groups. Figure 2.4 shows graphs of the four alternative uses of the turnout information in the data bank. It indicates that the best alternative was to use no turnout information at all. Although the correlation coefficients are the same in all cases, any use of turnout information increased the average error substantially by lowering the Democratic vote.[5]

Note, also, that omitting any turnout factor resulted in a slightly higher Catholic vote because working class groups, who are heavily Catholic, vote less regularly than the average. That in turn results in the best prediction being found at a slightly lower and more

[5] Besides the multiple weights on nonvoting used in the original simulation the simulations were also tried weighting every group's vote by $(1 - P_{35})$. The results were very much the same as for multiple weighting.

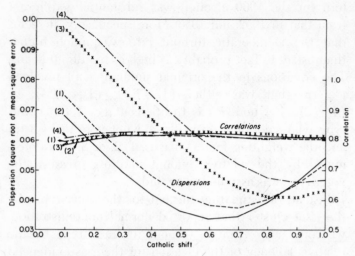

FIGURE 2.4 Effect of turnout corrections using data from 4-period master matrix.

(1) – – – – = Turnout factor (original simulation)
(2) ———— = No turnout or interest factor
(3) + + + = Interest factor
(4) —·—·— = Turnout and interest factor

reasonable C-shift, .6, than in the simulation with a turnout factor included, where it was .7.

Finally, if we use the 5-period data bank and omit any turnout correction, we get still better and more reasonable results (see Figure 2.5). With the usual correlation of over .8, we get a square root of the mean-square error of only about 3 percentage points with a Catholic shift of .5 or even with one of .4. We shall later use the latter combination of a .4 C-shift and no turnout correction in our "best-fit" simulation.

The failure of a turnout correction to improve the results is not a trivial finding. It is a finding in direct contradiction to a major social science theory that we and others have used and accepted for a good many years. It contradicts the proposition that persons under cross pressure vote less than others.

In the classic study by Berelson, Lazarsfeld, and McPhee, *Voting*,[6] to which the present study owes so much, it is asserted that nonvoting is one of the ways

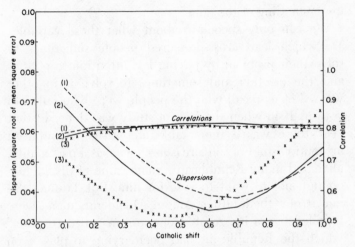

FIGURE 2.5 Effect of data-base and turnout corrections.
  (1) $---$ = Original simulation
  (2) ———— = No turnout correction; 4-period master matrix
  (3) $+++$ = No turnout correction; 5-period master matrix

that cross pressures become resolved. Persons faced with a dilemma about how to vote were found to become less interested in politics. They escaped the dilemma by leaving the field of political action.

That is a plausible theory. It fits general psychological propositions about human behavior. It was supported by the 1948 Elmira data on which the book,

[6] Bernard Berelson, Paul F. Lazarsfeld, and William N. McPhee, *Voting* (Chicago: University of Chicago Press, 1954).

Voting, is based. But it does not fit the 1960 national election data. We are forced to conclude that nonvoting is not necessarily a function of cross pressure. Sometimes it is and sometimes it is not. There are apparently some as yet unidentified intervening variables for which we must now search to restore some system to our understanding of the data.[7]

We can only speculate about what those variables are which lead cross-pressured people sometimes to solve their problem by paying less attention to politics and the election and sometimes to solve it in other ways. Let us recall who the people were who failed to vote in 1948 when the Elmira study was done. In that year there was a remarkable amount of Republican abstentionism. For some reason Dewey, as a personality, alienated many Republicans. Some came to dislike him enough not to want to vote for him. But Truman did not attract them either. Among Democrats many were repelled by an image of "Communism and corruption" which the Republicans had been trying to plaster on the Truman administration. They had an image of mink coats and Harry Dexter White lurking behind a little man in the White House. But Dewey did not attract them. In short, cross pressure in 1948 was alienation. Democrats driven from their party or Republicans from theirs had no place to go and no positive cause to serve by switching. Perhaps that is why cross pressure meant staying home.

[7] It must be mentioned to the credit of William McPhee, one of the authors of *Voting* and one of our colleagues in the initiation of the Simulmatics project, that in 1959 when we were setting up our model he expressed the view that his book had been wrong on this point. He warned against relying upon this mechanism in the model. Pool, on the basis of the published results, declined to be guided by McPhee's doubts and incorporated the nonvoting mechanism into the equations in accordance with cross-pressure theory.

In 1960 it was different. Protestant bigoted Democrats and Catholic Republicans were under cross pressure. But the pressures they felt suggested purposeful action. The bigot who felt that the country would be endangered by having a Catholic in the White House was much more politically motivated than the man who just lacked respect for both Dewey and Truman. If the bigot was to achieve his clear purpose, he had to vote. To a still greater degree the Republican Catholic under cross pressure would have reason to vote. He was not being pushed away from Nixon by dislike of him. On the contrary, he was being positively pulled toward something, whichever way he decided to vote. He was attracted by Nixon, and he was attracted by the image of Kennedy as a man of his own kind and by the need to protect his people against the attacks of the bigots. He was being pulled to a vote, not pushed from one.

Whether this difference between push and pull is what determines whether cross pressure leads to nonvoting or does not, we cannot say conclusively. We have been speculating about our data. But we can say with confidence that the simple theory that postulates nonvoting as the outcome of cross pressure can no longer be maintained.

In the jargon of long-standing psychological theories of conflict,[8] one would say that 1948 represented an "avoidance-avoidance" conflict for cross-pressured voters, whereas 1960 represented an "approach-approach" conflict. Conflict theory predicts that the subject in an "avoidance-avoidance" conflict will "leave

[8] Kurt Lewin, *A Dynamic Theory of Personality* (New York: McGraw-Hill Book Co., 1935), Chapter 4; Neal E. Miller, "Experimental Studies of Conflict," in J. McV. Hunt (Ed.) *Personality and the Behavior Disorders* (New York: Ronald Press, 1944).

the field," that is, avoid the choice if possible, whereas in an "approach-approach" conflict he will be motivated to "remain in the field," that is, actively consider both alternatives until a choice is finally made.

Continuing briefly in this realm of speculation, it seems clear that the 1960 Catholic Republican was faced with an approach-approach conflict, but the situation for the anti-Catholic Protestant Democrat is not really quite so clear. As noted above, a strong bigot would have found a Republican vote attractive. However, if the anti-Catholicism took the milder form of hesitancy rather than alarm at the prospect of a Catholic in the White House, then the emphasis would be on avoidance of a Kennedy vote rather than approach toward a Nixon vote. One alternative which we did not test would have been an asymmetric turnout correction that kept anti-Catholic Protestant Democrats away from the polls but propelled Catholic Republicans toward them. These lines of speculation are almost endless. At any rate, we can say with confidence that the simple theory that postulates nonvoting as the outcome of cross pressure can no longer be maintained. The whole question of nonvoting under cross pressure needs careful re-examination over a larger set of elections.

CONTINUITY PREDICTIONS

Now we have examined in detail the various parameters that went into the equations of our 1960 simulation. We have examined in particular the religious issue and the sensitivity of the results to assumptions about it.

But our 1960 simulation took into account only a

limited number of variables. There were many other and interesting variables that might have been tested out to see how well they could explain the November outcome. Past votes of various sorts, foreign policy, and civil rights all played a part in the campaign. It is to these other factors and their effects that we now turn.

Predictions based upon continuity of voters' attitudes from the 1956 Presidential election and from past Congressional voting, either using actual election statistics or the data bank do not give as valid predictions of the 1960 election as did the religious simulation.

The correlation between 1956 and 1960 returns in the real world is .43 for the 32 Northern states. Using the 1956 votes as indicated by the data bank the correlation is but .24.[9]

The correlation between 1958 Congressional returns and those in 1960 is .37. The correlation between the Congressional voting issue-cluster from the 4-period master matrix and 1960 is .53.

[9] The weakness of the data-bank correlation compared to the real-vote correlation reflects the imperfection of the poll returns as a measure of actual vote behavior. The state-by-state correlation between the 1956 election cluster in the data bank and the 1956 results is but .65, which is approximately the same level of correlation achieved in the summer of 1960 in predicting the 1960 election with last minute polls. The raw data-bank values, uncorrected by sensible estimates of biasing factors, are far from a perfect representation of real-world outcomes.

One interesting continuity was the fact that there was a .85 correlation for the 11 Southern states between 1956 and 1960. Although a correlation for 11 points is not very stable statistically, the South clearly is a special case and Congressional and Presidential voting are related there in ways quite different than elsewhere, as we shall see again later. Evidently what anti-Catholicism did in the South in 1960 was almost the same thing as what opposition to Stevensonian liberalism did to the South in 1956. The issues may have looked different, but the people whom these different symbols pried loose from traditional Democratic voting were substantially the same ones.

"LUCKY GUESS" PREDICTION

A "lucky guess" prediction derived from the 1956 Stevenson vote and the number of Catholics in each state would have achieved better results than the religious simulation if the appropriate guess could have been arrived at by some stroke of intuition. In that best lucky guess prediction, one simply takes the 1956 Stevenson percentage for each state and adds to it a certain fraction of the Catholic percentage of each state. As shown in Figure 2.6, a lucky guess of .35

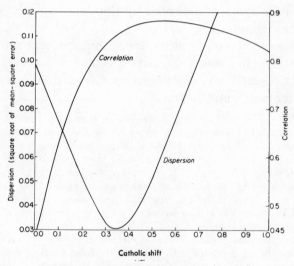

FIGURE 2.6 "Lucky guess" prediction.

would have given a square root of mean-square error equal to .03 and a correlation of .85 between the predicted and actual vote. Since the Democrats normally get about 60 per cent of the Catholic vote anyway, it would be extremely difficult to rationalize the appropriate guess as it would give 95 per cent of the Catholic vote to the Democrats. In short, the lucky guess is not so much a model of what happened as it is a post-factum discovery of the best set of numerical coincidences.

The numbers do, however, suggest something. Above we noted that in the South, the continuity from 1956 implies that Eisenhower, Stevenson, and anti-Catholicism were all epiphenomena; somehow each managed in its way to symbolize whatever more real loyalties and discontents were keeping some Southerners in the Democratic Party and driving some others out. So too, in the country as a whole it seems that aside from religious realignments, votes in 1960 and votes in 1956 followed much the same pattern. The Democrats whom the Ike charisma attracted and the ones whom the religious issue drove away were—barring the religious variable—people having the same social characteristics. The lucky guess model, though not a good model, sets us to asking to what extent the Ike charisma on the one hand and anti-Catholicism on the other were not also epiphenomena rather than causes and that they really only symbolized more determinative factors.

FOREIGN POLICY

A course that history might have taken in 1960, but did not, was one in which foreign-policy disputes came to affect the outcome of the election. We have recently simulated such a hypothetical scenario.

There turns out to be little relation between the actual outcome in November 1960 and the results obtained by simulating a campaign in which voters' decisions were dominated by foreign affairs.

In the environment of 1960 the issue of relations with the Communist bloc could have become the big question on the voters' minds. A Berlin crisis and/or Soviet peace offensive and/or a test ban treaty could have arrived on the scene to divide the Congress, the parties, and the American public. Suppose in 1960 the Democratic candidate had chosen to make a big pitch about achieving peace by negotiating outstanding political issues with the Russians. Suppose the Republicans had chosen as campaign themes that you can't do business with Bolsheviks, that we must roll them back, liberate the satellites, and not negotiate with insincere men. Suppose fighting in Laos or Vietnam had grown intense and Kennedy had said in his speeches that neither the Russians nor the Americans would gain from this; he, if elected, would negotiate a settlement. Suppose Nixon had said that only Quislings would settle for less than full freedom of those nations and had demanded that at any costs the Communists be defeated. That would have been a conceivable campaign quite different from the one that really happened, and it is the one that we have simulated recently.[10] It was, we must emphasize, not the campaign of 1960. Common sense tells us that. So does the fact that this simulation correlates with the actual results in November only +.43.

Yet the real-life campaign of 1960 was not devoid of foreign-policy issues. They just never became decisive, though at times it looked as though they might. In May

[10] The equations for the simulation used an issue-cluster called "negotiationism." It changed basic party vote for each voter-type according to how they distributed on that issue-cluster. That issue did not divide parties but did divide regions, sexes, religions, and races somewhat. The equation for the Kennedy percentage was: the average of the voter-type's Democratic Congressional vote and its 1956 Stevenson vote, plus one third of those scored as pro-negotiation and minus one third of those scored as anti-negotiation.

the U-2 crisis had occurred, dashing hopes of an evolving détente and plunging the cold war back into a phase of freeze. Early in the campaign Khrushchev announced that he was coming in person to the U.N. General Assembly Session. (The consequence that trip might have was one of the questions being pondered by campaign strategists in August.) His arrival and stampede through New York with its mixture of bluster, bombast, backslapping, and shoe thumping certainly focused some public attention on Soviet-American relations. In addition, Kennedy's assertion in the TV debates that the islands of Quemoy and Matsu were not necessarily objects to be retained from the Chinese communists at all costs injected the issue of a "hard" versus "soft" line further into the campaign. Nixon hopped on Kennedy's Quemoy comment, lambasting it as a dangerous concession. The issue was debated ad nauseam.

So the U.S. posture of "softness" or "hardness" in the cold war lurked always in the background of the campaign. It was an issue. But it never did break forth as one of those few issues that move many votes. It did not become a source of cross pressure.

That was how history worked out in 1960. Let us now put ourselves back in August and review how the prospective alternatives looked then before any one could know whether foreign affairs would generate crucial issues.

Our national telephone survey, conducted between August 13 and 18, asked voters what issues they thought were most important in the campaign. They were offered nine choices:

Keeping prices down
Government action to prevent unemployment

Helping old folks with their medical bills
Keeping up farmer incomes
Interference with states rights
Defending civil rights
Avoiding war by negotiation with the Russians
Keeping ahead of the Russians in production and science
Developing missiles to check the Russians

The order of presentation of these issues was rotated to prevent bias.

The result was that the three foreign-policy issues came out as most important, one, two, three. While the attitudes that the public expressed were moderate and pacific they were not working in Kennedy's favor. Kennedy at that time was widely believed to be young and inexperienced, especially in world affairs. Nixon

TABLE 2.1

OPINIONS ON COMPETENCE IN DEALING WITH THE RUSSIANS

Which Man Could Do the Best Job in Dealing with the Russians	Percentage of Kennedy Voters	Percentage of Undecided Voters	Percentage of Nixon Voters	Percentage of All Voters
Nixon	9	29	83	43
Kennedy	56	8	2	24
Both Equal	11	12	4	8
Don't Know	20	38	9	20
No Answer	4	13	2	5
	100	100	100	100

had been in world affairs. He had met Khrushchev in person and argued back to him. Only 56 per cent of Kennedy supporters in our poll asserted that he "could do the best job of dealing with the Russians." Eighty-four per cent of Nixon voters thought their man could do best. (See Table 2.1.) The voters saw in Kennedy's personal inexperience an instance of a long held conviction that the Republican party was better at keeping the peace than the Democrats. Democratic voters, of course, have more confidence in the Democratic Party and Republicans more in the Republican. The Independent voters are the significant ones here. Throughout the fifties they replied that the Democratic Party was better for preventing depression, the Republican better for keeping the peace. Which way they voted often hinged on which objective seemed more pressing. In our data bank 71 per cent of Independents said the Republican Party was better for keeping the peace and 29 per cent, the Democratic.

It is hard today when Kennedy, in death, has wrung rivers of tears from an admiring world to recall the frame of mind of August 1960. Without the record created by public opinion polls we might easily rewrite history through the prism of later events that made of Kennedy the foreign-policy leader of the world. But the fact is that in August 1960, voters had little sense that Kennedy had any competence in international affairs. Insofar as they focused on foreign affairs, they were more inclined to favor Nixon.

Indeed, this became clear from one other result of our August survey. The issue a respondent singled out as most important partly predicted how he would vote and also whether he had made up his mind. Those voters who singled out foreign policy as the most important issue in the campaign were most likely to have already made up their minds, and were for Nixon. (See Table 2.2.)

Incidentally, these people tended to be young, prosperous, and women. Forty-one per cent of women and only 29 per cent of men cited "Avoiding war by negotiation with Russian leaders" as the most important

TABLE 2.2

CANDIDATE PREFERENCE AS A FUNCTION OF
OPINIONS ON IMPORTANCE OF ISSUES

Most Important Issue	Percentage for Kennedy	Percentage for Nixon	Percentage Undecided	Total Percentage
Negotiation with Russians	37	44	19	100
Keeping Ahead of Russians	39	43	18	100
Developing Missiles	39	38	23	100
Keeping Prices Down	28	43	29	100
Civil Rights	36	38	26	100
Old Folks	51	20	29	100
Employment	52	18	30	100
States Rights	8	59	33	100
Farm Incomes	43	27	30	100

issue. That is one of several reasons why Nixon ran ahead of Kennedy among women. Our survey showed Nixon ahead of Kennedy 39½ per cent to 37½ per cent over-all, but with men for Kennedy and women for Nixon.[11] (See Table 2.3).

So much for the (alarming to Democrats) survey results on the role of foreign policy in the campaign. These survey findings sent us back to our data bank to look for more enlightenment. What further advice

[11] Other factors were operating too, to make Nixon in 1960 more appealing to women and the handsome young Kennedy less so—whatever press speculation may have said to the contrary.

This result is foreshadowed in the Simulmatics survey bank data. It shows that there was less anti-Nixon feeling among Democratic and Independent women than among such men:

Gap Between Independent Males and Females
Regarding Hostility to Nixon

Percentage Points

City	Men 11 over Women
Town	Men 6 over Women
Rural	Men 2 over Women

TABLE 2.3
VOTERS' PREFERENCE BY SEX, WHITES ONLY

	Percentage of Men	Percentage of Women
Kennedy for Sure	30	20
Leaning for Kennedy	12	13
Kennedy Total	42	33
Nixon for Sure	25	29
Leaning Nixon	12	14
Nixon Total	37	43
Undecided	21	23
Total	100	100

could be derived by looking back at the collection of old figures on public attitudes toward the various foreign-policy issue-clusters?

Examination of the data bank produced somewhat more comforting news for the Democrats. The data indicated that in the ten years just past, while foreign affairs concerned voters they did not provide incitements to switching votes. As we put it in one of our 1960 reports:

It can be seen . . . that foreign policy was important among both men and women.

. . . It is easy to say that an appealing foreign policy involves both strength and peace, but it is hard to implement this platitude in ways voters will understand.

. . . Groups do not know where they stand on many foreign policy questions. Finding the exceptional groups who have clear views is important, but it should be kept in mind that they are exceptions. . . .

. . . *Public interest in foreign affairs does not usually mean support of any particular policy.* With important exceptions, the public has not lined up on foreign affairs issues *in any ideological way*. What it asks for is foreign policy leadership in which it can have confidence—*not* particular foreign policies. . . .

Kennedy's problem is to demonstrate competence in foreign policy in ways that are, to use the Republican cliché, both peaceful and firm. The solution to this will not be found in the particular foreign policy positions taken on specific issues.

As of the moment it looks as though the election may turn on three key issues:

1. Party
2. Religion
3. Competence in foreign affairs

Civil rights, labor, jobs, etc. affect important special groups, but the above three issues seem vital across the nation. . . .

The Simulmatics model tells us what proportion of each type of voter supported Kennedy or Nixon in trial heats since 1958. . . .

In our analysis of the votes of *each* of the 480 voter-types, two things stand out clearly: *party lines and religion largely determined the votes*.[12]

In short, foreign affairs seemed clearly important, but if Kennedy demonstrated competence in them, he would not lose votes by virtue of his or Nixon's particular stands, so we believed.

While our report reached this conclusion, still we suggested doing a simulation of what would happen if contrary to our expectation votes were changed by foreign-policy issues. Specifically we wished to simulate what would happen if the hard-line, soft-line issue became so salient that it divided the voters and did affect their vote. That simulation was not done in 1960.

[12] Simulmatics Report No. 4, "Kennedy, Nixon, and Foreign Affairs," August 25, 1960, unpublished.

Now ex post facto we have run it. As noted above, it was not anything like the campaign of 1960, correlating only .43 with the November result. That is not even up to the .53 correlation between the basic party strength in each state as measured in four Congressional elections and the 1960 Presidential vote. But as an exercise in hypothetical history—what might have been—it is an interesting experiment.

In the simulated election contest Nixon would have won and won easily. However, it should be remembered that the absolute prediction is subject to the variable nature of the cutting points of items defining the issue-cluster. The cluster contains certain very unpopular items involving negotiations with Red China and some popular items on Big 3 meetings. For a prediction to be good in absolute value the balance of questions in the cluster must be appropriate to the balance of the issues debated in the campaign. The simulation gives Nixon 54 per cent of the national vote. It gives him every state outside the South and Kennedy 11 states in the South.

These results arise in the first place because this is a one-factor simulation. We are assuming out of existence the Negroes' struggle for civil rights and its impact both on the South and in the cities of the North. We are assuming out of existence the religious issue. We are assuming, in short, an election in which nothing but foreign policy affected basic political predispositions, and in which foreign-policy debate took the form of the Democrats being more conciliatory and the Republicans tougher. Remember also that the sharpest division on that issue is between men and women, and they are found in about the same proportions in all states. Remember also that we have found that in reality relatively few people are clearly committed on that issue. What most want is confidence in their leaders and to feel in general that their own party is competent. Remember finally that if we do postulate a situation in which people are shifted by this issue, whatever net gains there are accrue to the Republicans in the simulation, although this depends upon the cutting point factor discussed above.

What the simulation says is that in the South the Republican gains on this issue would have been smaller than were the Republican gains in the real election dominated as it was by religion and civil rights, items here wiped out. So the South, while shifting somewhat towards Nixon, would on balance have remained Democratic. Elsewhere the rather uniform Republican gains would have been enough to turn the election.

Let us look a little more closely state by state at what the simulation says would have been the difference between the result of a foreign-policy contest and the result of an average Congressional contest—which is a good measure of the basic party division.[13] We rank the states by the amount of effect that foreign affairs would have had in the simulated election—its effect is always in a pro-Republican direction. (See Table 2.4.)

A tough line appeals more strongly in the South than anywhere else, but does not all by itself change enough votes, considering how far back the Republicans start, to reverse the Southern outcome.

A tough line pays off next best in states that have substantial urban Catholic Democratic populations who are subject to cross pressure on the hard-soft issue. The effect is smaller where Republicans predominate anyhow and where therefore fewer people are available

[13] It might have been better to have used the average of Congressional and 1956 votes which we actually used as the base in this simulation, but the picture would be basically similar.

TABLE 2.4
RANKING OF STATES BY AMOUNT OF EFFECT OF FOREIGN-AFFAIRS SALIENCY*

State	Kennedy Vote 1960 (Actual)	Democratic Percentage in Congressional Vote-Cluster	Predicted Kennedy Percentage in the Foreign-Policy Simulation	Difference (i.e., Impact of Foreign-Policy Issue)
Oklahoma	.410	.755	.577	.178
Texas	.510	.740	.569	.171
Arkansas	.539	.737	.570	.167
North Carolina	.521	.732	.568	.163
Virginia	.473	.729	.566	.162
Florida	.485	.725	.565	.161
Georgia	.626	.721	.564	.157
Alabama	.577	.718	.563	.155
Louisiana	.638	.713	.559	.154
South Carolina	.512	.714	.562	.152
Mississippi	.596	.704	.558	.146
Delaware	.508	.595	.498	.097
Maryland	.538	.581	.487	.093
Kentucky	.464	.573	.483	.090
West Virginia	.526	.566	.479	.087
Rhode Island	.638	.514	.428	.086
Tennessee	.464	.559	.474	.085
Massachusetts	.605	.494	.416	.077
Connecticut	.537	.479	.410	.069
New York	.528	.506	.437	.068
New Jersey	.504	.481	.414	.067
New Mexico	.502	.514	.450	.064
California	.497	.521	.460	.061
Arizona	.445	.508	.447	.061
Illinois	.501	.516	.457	.059
Colorado	.458	.497	.441	.056
Michigan	.511	.503	.447	.055
Montana	.487	.486	.430	.055
Ohio	.468	.490	.438	.051
Washington	.491	.485	.434	.051
Oregon	.475	.479	.428	.051
Wyoming	.449	.476	.425	.051

(*Continued*)

TABLE 2.4 (*Continued*)

State	Kennedy Vote 1960 (Actual)	Democratic Percentage in Congressional Vote-Cluster	Predicted Kennedy Percentage in the Foreign-Policy Simulation	Difference (i.e., Impact of Foreign-Policy Issue)
Wisconsin	.481	.488	.439	.050
Missouri	.507	.481	.432	.049
Nevada	.512	.480	.432	.048
Minnesota	.507	.478	.430	.048
Pennsylvania	.513	.436	.390	.047
Idaho	.464	.457	.414	.043
Utah	.452	.468	.425	.043
Indiana	.448	.461	.419	.042
Nebraska	.385	.457	.416	.041
Kansas	.393	.452	.413	.039
Iowa	.431	.449	.411	.038
North Dakota	.445	.450	.412	.038
South Dakota	.417	.445	.409	.036
New Hampshire	.466	.402	.367	.035
Vermont	.414	.380	.357	.023
Maine	.430	.369	.349	.020

* Note that here and throughout this section we use not the *actual* Congressional vote as it was cast, but the public opinion poll replies about Congressional vote as it appears in our data bank and as it arises when we make up synthetic states. Our purpose here is to test a methodology that can be applied to individual voter-types and to groups of them. We would gain in solidity but lose in learning about the potential of our method if we chose some specific Congressional election outcomes as our criterion.

to shift. The Midwest shows no distinctive pattern in this simulation. The urbanized states in the Midwest behave like other states with substantial urban Catholic Democratic populations. The solidly Republican states of the great plains behave like solidly Republican states elsewhere.

Before we leave the area of foreign affairs let us

review what else our data bank could have told us. The simulation we ran used just one foreign-affairs issue-cluster—the one we called negotiationism. We list in Table 2.5 the clusters which yielded interesting

TABLE 2.5

REGIONAL DISTRIBUTION OF FOREIGN-POLICY ATTITUDES

Issue	South	Border	East	Midwest	West
Pro-Negotiations	37%	38%	43%	44%	45%
Show Foreign-Policy Knowledge	52	59	71	69	73
Sophistication on Foreign Policy	42	37	55	50	53
Favor Committing Troops Abroad	42	45	53	45	51
Defend Europe	25	26	31	34	31
Sympathetic to Europe	31	28	46	36	41
Soft Policy to Red China	20	24	26	26	28
Favorable to Asian Neutrals	40	30	50	49	43
Disarm H-Bomb	28	24	33	32	31
Pro-Preparedness	31	36	39	34	39
Pro-United Nations	60	61	67	67	67

results when responses in the various regions were compared. The big differences are between the South and the rest of the country. It is the South, not the Midwest, that is least internationalist, as that concept has come to be defined in the past twenty years. The people of the South, and of the Border states too, are least informed about foreign affairs, least interested in Europe, least understanding of Asian neutrals, least interested in disarmament or the United Nations. Midwesterners join them only on the issues that involve the possibilities of commitment of American troops abroad, and on armament spending, not on the more purely political issues. The East is distinctive only in its sympathetic attitudes toward the problems and policies of Europe. But by and large, regional differences are small, though for the South unambiguous.

Party differences are even smaller. That is why

foreign-policy issues had so little impact on the election outcome. Even with Khrushchev storming through New York in the middle of the campaign, and with Kennedy and Nixon arguing about Quemoy and Matsu, most people in 1960, though thinking foreign policy the most important issue, found in foreign policy little reason to question their normal voting inclinations.

CIVIL RIGHTS

Another issue that we explored by a simulation after the fact was civil rights. The first report we prepared for the 1960 campaign was one on the Negro vote. It contained a review of the relevant facts in the data bank. Only after the election did we use the data to simulate a hypothetical campaign in which the only vote determinant other than past voting habits was attitude toward civil rights. And once again we find that this was not an issue that caused any significant amount of cross pressure except in small groups of voters.

It was a terribly important issue for minorities of voters such as Negroes or segregationists. But it could not account for the major shifts among the bulk of the voters.

How important "Negro Voters in Northern Cities" might be we indicated in our very first report in June 1960. (We quote the report at some length to give the flavor of these reports.)

In a close election even a moderate shift in Negro votes could be decisive in eight [key Northern states]. As long as the over-all division of votes is less one-sided than 53–47, any shifts in the Negro vote could determine the outcome.

Note: The validity of this conclusion is sometimes denied on the grounds that many Negroes are non-voters. If that were true, obviously their weight in the electorate would be proportionally less than their numbers would indicate. However, it is not true that Northern Negroes stay away from the polls. This belief is based upon national public opinion poll figures which combine figures from both North and South to produce enough cases for separate analyses. The Simulmatics procedure of combining polls in a single treatment permits us for the first time to make separate comparisons of non-voting:

> Between middle class Negroes and
> whites in Northern cities;
> Between lower status Negroes and
> whites in Northern cities.

The conclusion is that non-voting is *not* significantly disproportionate between Negroes and comparable whites in the North, and that if there is any difference it is that Negroes vote more. It is true that lower class people vote less than higher class ones, so that there is some validity to the argument that the Negro voting potential is reduced by the larger proportion of Negroes in the lower income groups. But the effect of this on Negro turnout is not very great since class by class there is no difference in Negro and white turnout.

In the past the Northern Negro vote has been a reservoir of Democratic strength.

In 1952 the split in the North was:

	Negro voters*	All voters†
Stevenson	74%	43%
Eisenhower	26%	57%

In short, the Negro vote in the Northern cities has been 3 to 1 Democratic.

What has happened to the Negro vote since 1952?

* Based on 1,494 public opinion poll interviews summarized in the Simulmatics computer program.

† Based on official returns.

In 1952 and 1954 the Democrats still held their lead in Northern Negro votes.

But in 1956 and 1958 they lost a small but significant number of them to the Republicans.

Here are the figures on *party preference** as revealed in 4,050 poll interviews with Northern Negroes:

	1952	1954	1956	1958
Democrats	58%	59%	53%	52%
Republicans	18	17	23	24
Independents	24	25	24	24
	100%	100%	100%	100%

Note some interesting differences in rates of defection.†

a) One fact which makes this as yet small defection to the Republicans of even greater importance is that it appears to be stronger among middle class Negroes, i.e., among those who are molders of opinion, such as journalists, ministers, professional men.

As to party affiliation we find that about 20% of middle class Northern Negroes called themselves Republicans in the early part of the Eisenhower era. By 1958 that figure was up to 31%.

b) The shift of Negro Independents to the Republicans reached striking proportions, particularly in the East, after 1956. Our sample showed that in 1956, 70% of Negro Independents in the Northeast had voted for Eisenhower. The previous vote was not more than 25%. In 1958, about 60% of Negro Independents in the Northeast voted Republican for Congress. . . .

* Party Preference figures (as distinct from candidate vote) are used here to compare Presidential and off years on the same basis.

† It should be taken into account that these results are derived from small numbers. Middle class Negroes in the North are only about $\frac{1}{10}$ of the Negro population. Middle class Northern Negroes interviewed in 1958 numbered 116, a probable but not indisputable basis for reaching the conclusion indicated.

The shift was not just a swing to "Ike."

It was definitely a shift in party loyalty. It will *not* be recovered with "Ike" out of the running.

What evidence supports this proposition—which we regard as the most important in this report?

The evidence comes from two sources:

Comparison of Presidential and Congressional votes;
Poll questions about the parties as such.

a) The trend of Negro voters toward Eisenhower in 1956 was matched by the same trend of the white population. In 1952 Eisenhower got 55.4% of the popular vote. In 1956 he got 57.7%, a gain of 2.3 percentage points.

But the Negro shift and the white shift were of a very different character. Among comparable groups of white non-Republicans the Eisenhower vote was much larger than the vote for Republican candidates for Congress. Among Negroes that was *not so*. They were almost as likely to swing Republican for Congress as they were to swing to "Ike."

In short, the Democratic affiliation fell and the Republican affiliation grew by about 7 percentage points.

b) Besides this loss through shift in party affiliation, there was a further Democratic loss in 1956 when Independent Negro voters supported Republican candidates in greater numbers than ever before.

In 1952 about 25% of the Northern Negro Independent vote went to Eisenhower. In 1956 this had risen to about 45%. The Independents, it will be recalled, are about ¼ of all Northern Negroes (as they are of the whites). It is among the Independents that the defection from the Democrats has been greatest.

The following table of public opinion poll results among Independent voters gives the Republican percentage of the Congressional vote subtracted from the Republican percentage of the Presidential vote. For example, in 1952, in the Midwest 29% of the lower status

Negroes voted for "Ike"; 25% voted Republican for Congress. The difference is 4 percentage points in "Ike's" favor—the number entered in the table. The bigger the number the more splitting there was to vote for "Ike" but not for the party.

PERCENTAGE FOR "IKE" MINUS PERCENTAGE
REPUBLICAN FOR CONGRESS

		1952		1956	
		Negroes	Comparable Whites	Negroes	Comparable Whites
East	Middle Class	0	12	3	6
	Lower	0	13	4	14
Midwest	Middle Class	−4	28	0	25
	Lower	4	26	−11	9

Note that among all groups of white Independents there was a substantial personal following for "Ike" which did not carry over to Congress. *Not so among Negro Independents. . . .*

The Negro Republicans are no longer so torn. Previously, about one in six, although himself a Republican, thought the Democratic Party better for his people. Now only one in twelve think so.

The Independents no longer see the Democratic Party as a better friend of the Negro than the Republicans. They are beginning to see them as alike as Tweedledum and Tweedledee.

Note that one cannot simply subtract the numbers from the above tables to obtain the percentage of those preferring the Republican Party, for there is a third possible response: to make no choice on the grounds that as far as the respondent is concerned the parties are identical. The following table shows the percentage of Independents who chose neither the Republican nor the Democratic Party as best for their people.

INDEPENDENTS: PERCENTAGE CHOOSING EACH PARTY AS BETTER FOR PEOPLE LIKE THEM

	Respondent Chooses Republican Party	Democratic Party	Refuses to Choose	
1952	12	47	41	100
1956	20	29	51	100

Our conclusion must be: the Democratic Party has lost some of its appeal. The Republican Party has gained some. But there has also been general disenchantment with both parties. A growing group of Negro voters do not expect much from either right now.[14]

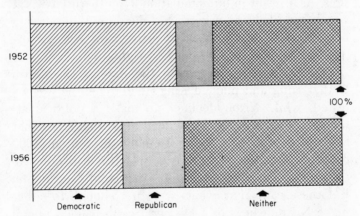

Negro Independents: Which party is better for people like them?

Thus the report documented a problem and an opportunity. Kennedy was in a position to pick up an extra percentage point, or half of one, in each major

[14] Simulmatics Report No. 1, "Negro Voters in Northern Cities," May 15, 1960, unpublished.

industrial state by a strong civil rights stand. However, a civil rights simulation fails to reproduce the voting shifts that actually occurred in 1960 since its effect is so pinpointed. The correlation of our civil rights simulation with the actual November results was only .41 for the 32 Northern states and .47 for all 48 states.

In part the weak impact of Kennedy's forceful espousal of civil rights on the pattern of votes is attributed to his opponent Nixon. Some Republicans might have captured the South while alienating the Negroes by making a clear-cut differentiation between their stand and Kennedy's. Nixon did not do that; he too took a fairly strong, though not equally strong stand for Negro rights. We were simulating a Kennedy-Nixon campaign, not a campaign between polar opposites. As a result in our simulation a Southern Dixiecrat was more likely to express his discontent by staying home than he was by switching to Nixon who was expressing himself in favor of civil rights too.

True, Kennedy did rise to the defense of Martin Luther King and phoned King's wife when he was arrested, while Nixon declined to make a political attempt to influence a judicial decision in that case. That helped to push the Negro Democratic vote in 1960 back up from the two thirds to which it had fallen to something like three quarters of the total Negro vote. But these campaign events were not enough to either make Nixon a hero to a Dixiecrat or to make a liberal Republican turn against him, for Nixon was a civil rights advocate too.

In short, his strong civil rights stand may have won a few key Northern states for Kennedy by its impact on Negro voters. But it had less impact than it might have had between polar candidates, and whatever affect it had was not general enough in the population to make the distribution of civil rights views a good predictor of state-by-state results.[15]

Race and issues of race did indeed make the voting results in the South differ from those in the rest of the country. (That we shall discuss in the next chapter.) But racial feelings, like world affairs, were to the candidates more a potential source of trouble waiting in the wings, than determinants of much of the final vote. They were, however, an added factor that created significant waves that were superimposed upon the great swells of opinion caused by other variables.

[15] The simulation that we used separated the civil rights saliency from civil rights attitude-clusters. It caused those individuals for whom civil rights were very salient to shift in directions determined by their attitude.

3

The Dynamics of 1960

THE USES OF SIMULATION

Simulation is a new tool. It has its precedents, but it is novel enough in the kit of science to need some comment on its potential uses.

The philosophy of simulators is not identical with that of other researchers. Simulators explore worlds that might be. Their more empirical colleagues establish facts about the world that is. The simulator brings the rigor of science to bear on a range of fascinations that until now have been the province of the writer of fiction, the essayist, or the mystic: the "what if" question of history. Such are the questions to which the poet addressed himself when he asked what would have happened "for want of a nail." They are queries about worlds that never were but might have been.

The philosophical status of such "what if" questions is not clear. What exactly does it mean to ask what might have happened in history had no illegal ballots been cast in Illinois in 1960—which some Republicans allege would have reversed the result of the election.

Or what might have happened if Lee Oswald or John Wilkes Booth had aimed 3 inches astray. The dogmatic empiricist rejects such questions. In a determinist world what happens is in the last analysis what had to happen: questions about other paths of causality may be dismissed by the empiricist as metaphysical. A less dogmatic empiricist might concede that there is some mystery to free will or indeterminacy but leave that mystery to the philosophers. As scientist he explores the facts and he works with a model of strict causality. There is, however, something unsatisfying about a set of canons that precludes the scientist from addressing himself to the "what if" questions that are the constant concern of prudent men.

Simulation is a way of analyzing such fanciful questions scientifically. One sets up a model of the hypothetical world one wishes to analyze, with all its relevant complexities. The model can then be manipulated with one or more of several alternative purposes: prediction, postdiction for analysis, exploration of alternatives for sensitivity testing, exploration of policy alternatives.

We began this essay by reporting an experiment in prediction. In 1960, three months before the election, we simulated November voter behavior, given the world as it seemed likely to be at that moment. The test of that simulation's success was its match to what actually happened three months later. We applied that test in Chapter 1.

In Chapter 2 we used simulation in a different way —sensitivity testing. Having shown in Chapter 1 that our model could predict tolerably well, we used it to vary, in imagination, a number of parameters of the situation, and to determine how much difference they made. We varied the intensity of anti-Catholicism, and of Catholic loyalty to a Catholic candidate. We also varied turnout patterns.

We varied some voter attitudes which had no great bearing in 1960 but which could have been significant in worlds that might have been. The election might have been a very different one had it turned out to be a policy contest between a clear-cut hard and clear-cut soft line on foreign affairs. It was not that nor anything near it, yet we can simulate with fair confidence and substantial detail what the outcome of such a contest might have been. We have a model sufficiently detailed, sufficiently specific in its postulations, and apparently sufficiently accurate to permit us to believe that we can work out the main interactions in such an imaginary world.

In the present chapter, we shall use simulation in still another way—postdiction to analyze what actually did happen in 1960. We shall try to improve the model we used until it corresponds as closely as it can to reality. Then what goes on inside the model may be suspected to represent what was going on in reality.

There is, of course, no trick to figuring out in retrospect a set of parameter values that gives a good prediction. The point of doing so is not to crow about how good the match is, for we pick the parameters to achieve that match. The point to finding the "best" predictive model is that an analysis of a model that matches the real world may indicate what was happening in the real world.

We may by postdiction estimate, for example, what the extent of anti-Catholicism was, or what proportion of Protestant Democrats swung over to Nixon. We may estimate how far the verbal answers to public opinion polls were an accurate statement, or an understatement, of prejudice. The amount of prejudice postulated in the model that best fits the results is likely to be a good estimate of the amount of prejudice in the real world.

Note, however, that we have said only that the values in a good model *may* or *are likely to* provide good estimates. A good match between various outputs of a model and reality is not proof that the inner workings of the model are the same as the inner workings of the real-world system that the model imitates. As every social scientist knows, correlations can be coincidental. A representation of human eating habits that made consumption of corn flakes a function of the position of the sun would predict well, but would not well represent the psychological and social mechanisms at work in the real world.

If we are about to examine the "best-fit" simulation of the 1960 election it is only because it is highly likely to be an image of reality, not because it can be proved to be one.

That raises a crucial question for any simulation study: How can the model it uses ever be validated? Validation is a matter of degree. A complex model can never be validated by any single observation. The more independent variables that there are in a model the more dependent variables the model must predict before it can inspire confidence. And the more items it predicts correctly the greater the likelihood that the model is in a 1-to-1 relation to reality.

For example, it is no test of an election model that it happens to pick the winner of a Presidential contest. A number of methods ranging from following the gambling odds to sheer luck have at least a 50-50 chance of doing that. It is a much more rigorous test of a model to require that it produce a prediction for a distribution over 48 states. If it passes that test, the chance that the model is only coincidentally related to reality is relatively small, though it still could be. Additional tests would be whether the model correctly predicts men versus women, Negroes versus whites, rich versus poor, and so on. With every successive test of correspondence to reality that a model passes, we gain confidence that its dynamics are those of the real world.

That is why we are interested in postdicting with the best rational model. By a "rational" model we mean one using mechanisms that it is plausible to believe are the real ones (for example, cornflakes consumption being related to people's habits, not to the position of the sun). A rational model that can postdict many facts is suggestive of what the real facts may be. And so we took the model we used in 1960, shifted the parameters to pick the best-fit values, examined where the predictions were still poor, introduced other variables, and wound up with a model about as good as the variables and data available to us would permit.

THE BEST-FIT SIMULATION

This best-fit simulation used exactly the same equations as our 1960 one with the following exceptions:

1. We dropped the turnout correction that had added nothing to the accuracy of our prediction.

2. We used the 5-period instead of the 4-period master deck of data.

3. We assumed a shift to Kennedy of .4 of Catholics who otherwise would have voted Republican. A .3 shift is indeed almost as good and better by some criteria, so this slight departure from the .33 used in 1960 is not really significant.

4. We introduced a new shift in the Southern states only, namely a postulate that 10 per cent of Negroes who would

otherwise have voted for Nixon voted for Kennedy, and 10 per cent of whites who would otherwise have voted for Kennedy voted for Nixon.

The effect of the last correction was to recognize that our previous simulation overstated Kennedy strength in the South. There was more working against Kennedy in the South than just anti-Catholicism. States rights, civil rights, conservatism, and race were adding to white Democratic defections, partially offset by the small but growing Southern Negro vote.

In August 1960 we had declined to predict the South by use of the religious issue simulation. It was clear to us then that a variety of other factors were also working there. We had said in our report:

In this particular simulation, results for the South could be misleading. A more complicated analysis of the South can be made in which the organizational situation in each state (e.g., the status of pledged electors, one-party tickets, etc.) is a variable of the simulation. So much in the South depends on what a few leaders do that a meaningful *state-by-state* simulation would have to include these variables.[1]

As we shall see, that was exactly right. No over-all social factors could bring into line prediction of the Southern states, one by one. State machine decisions produced major variations. But the South as a whole ended up just under 10 per cent less Democratic than party loyalty and the religious issue alone would have led us to predict.

Table 3.1 presents the results of postdiction with the best-fit simulation. The predicted and actual vote in each of the 48 continental states is given.

[1] Simulmatics Report No. 2, "Kennedy Before Labor Day," August 25, 1960, unpublished.

TABLE 3.1
Best-Fit Simulation

State	Predicted	Actual
Louisiana	.66	.64
Texas	.61	.51
Rhode Island	.59	.64
Oklahoma	.57	.41
Delaware	.57	.51
Florida	.57	.48
Arkansas	.56	.54
Virginia	.56	.47
Alabama	.56	.58
Massachusetts	.56	.60
North Carolina	.56	.52
Mississippi	.56	.60
South Carolina	.55	.51
Georgia	.55	.63
Maryland	.55	.54
New Mexico	.54	.50
New York	.54	.53
Connecticut	.54	.54
Kentucky	.52	.46
California	.52	.50
New Jersey	.52	.50
Illinois	.51	.50
West Virginia	.51	.53
Arizona	.51	.45
Tennessee	.50	.46
Wisconsin	.49	.48
Michigan	.49	.51
Colorado	.48	.46
Montana	.47	.49
Ohio	.47	.47
Minnesota	.47	.51
Pennsylvania	.46	.51
Washington	.46	.49
Nevada	.46	.51
Missouri	.45	.51
Wyoming	.45	.45
New Hampshire	.45	.47
Oregon	.45	.47
North Dakota	.44	.45
Nebraska	.43	.38

(*Continued*)

TABLE 3.1 (*Continued*)

State	Predicted	Actual
Indiana	.43	.45
South Dakota	.43	.42
Vermont	.42	.41
Utah	.42	.45
Iowa	.42	.43
Kansas	.42	.39
Idaho	.41	.46
Maine	.39	.43

How close is that postdiction?

For the country as a whole it gives Kennedy about the vote he got. It gives him 322 electoral votes which is 8 more than what he actually won on the continent. The mean deviation of the postdicted Democratic percentage of the party vote from the actual is but $9/10$ of 1 per cent in the 32 states of the North and but $6/10$ of 1 per cent for the country as a whole.

But to pick parameters retrospectively such that they fit the results for the country as a whole is easy. The first true test of the model is whether it places the individual states in their proper relationship. One test of that is the coefficient of correlation across states between the postdiction and the November results. That is the test we dared accept in 1960. The modified best-fit simulation that we are now using does no better on the Northern states than our 1960 simulation from which it differs only trivially, but it brings the Southern states into line to some degree.

Product Moment Coefficient of
Correlation Across States Between
Simulation and November Outcome

32 Northern States	.81
48 States	.70

The correlation tells us that about two thirds of the variation in election results between states in the North and about one half of the variation in election results across the country are explained by the model used in this best-fit simulation. There were indeed other factors operating such as local traditions, variations in the effectiveness of party organization, local leadership, the views of the local press, and so on. These, and random variations, however, accounted for but one third of the variance in Northern results, although if we include the South, with its more parochial and rooted party system, they account for one half of the variance.

A still more rigorous test of postdiction can be applied, namely absolute deviation of the simulation from the election results. (Note, for example, that if we were exactly 5 percentage points high in every single state, the coefficient of correlation would be 1.0, or perfect, while the average deviation would be 5 percentage points.) This is the criterion we declined to try to satisfy in 1960, but which in retrospect we see was achievable. Let us examine our best-fit simulation by that criterion.

The median absolute deviation was, as we can see, reasonably small and, as expectable, smaller in the North.

Median Percentage Points of Difference
Between Kennedy Vote in Simulation
and in Election

32 Northern States	2.1%
48 States	2.5%

Perhaps a better measure of error is the square root of the mean squared deviation of actual from simulated results. That yields essentially the same picture.

Root Mean Square of Differences
Between Percentage for Kennedy in
Simulation and Election

32 Northern States	3.2%
48 States	4.7%

The model reproduces enough of what was different about states to account fairly well for results state by state.

Thanks to the Gallup Poll, a few more tests can be applied to the simulation. Did it postdict the votes of men versus women correctly, of Negroes versus whites, of Catholics versus Protestants, of upper- versus lower-income people?

Such comparisons cannot be made between the simulation and actual November returns because the actual vote is not recorded by voter-type. Comparison can be made between the simulation (based as it is on poll figures only up to and including 1959) and Gallup Poll figures from the actual election period a year later in the Fall of 1960. Since both these measures are random variables, differences may occur either because the simulation is off or because even the late poll results report voter intentions with some margin of error, or both.

The late Gallup Poll results are a good measure of group votes. The late October–early November Gallup surveys predicted the election with an over-all across-the-board error of but ½ per cent. This is not really "prediction," but rather reporting. It is not prediction because the poll does not attempt to anticipate how people will change. Rather, the late surveys come after the voters have made up their minds. Those surveys report voter decisions. Whatever error arises is because the report is obtained on a sample, rather than on a total population basis. No responsible poller claims to do other than report the state of opinion as of the date of his interviews in the field. In 1960 the Gallup Poll was probably accurate both at the beginning of August, when it reported Nixon 6 points in the lead, and in October–November at which time the over-all poll figure was able to be checked against the real vote. But neither figure was strictly speaking a prediction.

TABLE 3.2
KENNEDY PERCENTAGE OF THE TWO-PARTY VOTE

	As Postdicted by Best-Fit Simulation	As Reported by Gallup Poll*
Men	52%	52%
Women	49	49
Nonwhite	66	68
Whites	51	49
Catholics	80	78†
Protestants	39	38†
Jews	71	81
Metropolitan	58	60
Town	48	45
Rural	47	46
C; Manual Workers	61	60
A + B	44	43
Republican	14	5
Democrat	80	84
Independent	48	43

* *The Boston Globe*, January 27, 1964. The fact that the largest deviations are in party affiliation is interesting and revealing. People do not change their sex, or domicile, or occupation when they decide how to vote. They may, however, change their conception of their party affiliation to bring it into consonance with their vote intention. Apparently some persons did so, which may account for the very small numbers of Republicans for Kennedy and Democrats for Nixon in the Gallup Poll results.

† An independent set of poll results reported by the Survey Research Center in an unpublished paper found the Kennedy vote to have been 81 per cent among Catholics, 37 per cent among Protestants, confirming the agreement between the Gallup Poll and our simulation.

With some confidence, therefore, we use the late Gallup Poll results about how major social groups voted as a reasonable estimate of the facts. Our simulation, if it was making our 480 voter-types behave in realistic ways, should conform reasonably well to the Gallup estimates of the social division of the vote. Table 3.2 makes the comparison.

We have now tested our model against a variety of checks, and we know where it fits and where it does not. The purpose of all of this is analysis, not accuracy of prediction. What does the close fit at many points and the deviations on a few tell us about the apparent dynamics of voter behavior in 1960? What does the way our microcosm works tell us about the way the macrocosm worked?

SOME CONCLUSIONS ABOUT 1960

1. The Role of Political Habit

Most Americans do not know even the name of their Congressman.[2] They vote in Congressional elections primarily for party. If they swing from one election of Representatives to another, it is usually a reflection of support or opposition to a party or administration, not of support or opposition to a candidate.

If there are exceptions, and there are, they tend to cancel out between the parties. In one district an incumbent Democrat has made many friends over the years and draws votes for himself as an individual. In

[2] Donald E. Stokes and Warren E. Miller, "Party Government and the Saliency of Congress," *Public Opinion Quarterly*, Vol. 26 (1962), pp. 531–546. See also William A. Glaser, "Fluctuations in Turnout," in William N. McPhee and William Glaser, *Public Opinion and Congressional Elections* (New York: The Free Press of Glencoe, 1962).

another place a Republican does the same. The net result is that nationally the Congressional vote reflects the basic party division, and, since Democrats far outnumber Republicans, the majority is generally Democratic. Oscillations in party division from off-year to off-year are relatively small, reflecting, in the small changes, over-all contentment or discontent with the Administration.

In Presidential years the charisma of the man supervenes over party tradition. Millions of voters leave their party to vote for the Presidential candidate whom they believe to be "the better man." And as they do so they sometimes vote a straight ticket, enabling some Congressmen to ride the coattails of the winning Presidential candidate.[3]

In short, off-year elections reflect the basic party division closely and fairly stably from year to year. Presidential year Congressional elections show more swing. And Presidential elections themselves show still more swing.[4]

Our job is to explain a Presidential election. One factor is clearly party affiliation, which provides a base line of the vote, with whatever swing there is centering about it. In 1960, as in any year, party loyalty and party habit, measured by past Congressional vote, was the most important factor in determining the vote. Past Congressional vote was the main base which we

[3] For a recent discussion of the "coattails" phenomenon, see John M. Meyer, "A Reformulation of the 'Coattails' Problem," in William N. McPhee and William Glaser, *Public Opinion and Congressional Elections* (New York: The Free Press of Glencoe, 1962).

[4] An interesting analysis of such swings which explains the typical pattern of Congressional loss in off-year elections by the party of the incumbent President has been given in Angus Campbell, "Surge and Decline: a Study of Electoral Change," *Public Opinion Quarterly*, Vol. 24 (1960), pp. 398–418.

used in predicting the vote, with any other factors leading merely to deviations from that base.[5]

2. Anti-Catholicism

Millions of Protestants and other non-Catholics who would otherwise have voted Democratic could not bring themselves to vote for a Catholic. In total—so our model says—roughly one out of five Protestant Democrats or Protestant Independents who would otherwise have voted Democratic bolted because of the religious issue. The actual number of bolters varied with the voter-type and was determined in the model by the proportion of that voter-type who had replied on surveys that they would not want to vote for a Catholic for President.

What our model tends to show is that the poll question was a good one. The model suggests that the number of people who overcame the social inhibitions to admitting prejudice to a polltaker was about the same as the number who overcame the political inhibitions to bolting their party for reasons of bias.

The proportion in our data bank (among Protestant voters only) who expressed reluctance to vote for a Catholic varied with party and other variables as indicated in Table 3.3. Forty-nine and nine-tenths per cent of all Protestants nationally and 59.2 per cent in the 32 states of the North would have voted Republican anyhow based on their normal, that is, Congressional voting habit. But among the rest, who would otherwise have voted Democratic, a net loss of around one fifth was suffered by Kennedy because of religion. The apparent gross number of Democratic defectors who

[5] See Chapter 1 for a more precise statement of the past vote used for different voter-types.

thought religion was a reason for their action was still larger, but some of these voters would have defected anyhow on other grounds, such as general conservatism. They had done so already in elections in which the religious issue did not arise. Other defectors among

TABLE 3.3

PERCENTAGE OF PROTESTANTS WHO OBJECTED TO
VOTING FOR A CATHOLIC FOR PRESIDENT

Rural	34.8%
Town	34.8%
Urban	31.8%
A + B SES	36.1%
C SES	30.9%
Democrats	25.6%
Republicans	46.7%
Independents	30.0%
Total	34.4%

Democrats would be offset by the return to the party of occasional Protestant Democrats who had been voting Republican. But these were few. In any case, the net real loss by Kennedy seems to have been about one fifth of otherwise Democratic-voting Protestants.

3. Pro-Catholicism

Offsetting the loss of Protestant votes for Kennedy was a strong swing to him by Catholics. There had been a massive defection to Eisenhower and against Stevenson by Catholics in 1956. Fully 35 per cent of Catholic Democrats voted Republican in that year. With Eisenhower out of the running, any Democratic candidate but Stevenson could count on recovering some of these defectors. Kennedy did recover these Catholic defectors from the Democratic Party and more. Taking Congressional voting as a base for estimating normal party vote, over one third of Catholics who would otherwise have voted Republican seemed to switch to Kennedy. The best guess is around 40 per cent. There are two and a half times as many Protestants as Catholics in this country, but the Catholic shift was more pronounced. If our model is right, the Protestants who shifted because of the religious issue outnumber the Catholics who did so by 1,500,000. Kennedy gained 2,800,000, and lost 4,300,000 votes. Seven million voters out of 68,000,000 defected because of religion. [*]

There has been much debate as to which candidate gained by the religious issue. Some analysts have argued as we do that more Protestants shifted than Catholics, but others have said the reverse.

The simulation gives an interesting answer to this debate. The shift of one in ten American voters on religious grounds cost Kennedy one and a half million votes, or 2.3 per cent of the total vote. But while Kennedy lost in popular vote he gained in electoral votes on the religious issue. The best-fit simulation indicates that Kennedy netted 22 electoral votes because of the religious issue! Table 3.4 reveals that bunching of the Catholic shift in large, closely fought, industrial states, and the location of much of the Protestant shift in "safe" Southern states gave Kennedy this net advantage despite a popular vote disadvantage. By our calculations, Kennedy lost by the religious issue the following states he otherwise would have won:

[*] Using polls, the Survey Research Center estimated that 10.8 per cent of the voters, 7,350,000, switched on the religious issue, Kennedy's net loss being 2.2 per cent, i.e., 1,500,000. Philip E. Converse, Angus Campbell, Warren E. Miller, and Donald E. Stokes, "Stability and Change in 1960: A Reinstating Election," *American Political Science Review,* Vol. LV, No. 2 (June 1961), p. 278. This is a remarkable concurrence of results by different methods, with different data, from different periods, S.R.C's from the campaign period itself, ours from years before.

	Electoral Votes
Kentucky	10
Tennessee	11
Florida	10
Virginia	12
Oklahoma	8
Montana	4
Idaho	4
Utah	4
California	32
Oregon	6
Washington	9
	110

He won the following states he would otherwise have lost:

Connecticut	8
New York	45
New Jersey	16
Pennsylvania	32
Illinois	27
New Mexico	4
	132

4. Civil Rights, Foreign Policy, and Other Issues

No issue other than religion had a significant nationwide net effect on the vote. We noted but dismissed in the last chapter two other issues as major determinants of the outcome. Neither foreign-policy attitudes nor attitudes on civil rights correlated well with the November results.

In the case of civil rights, some portions of the population were significantly affected. We described these effects as waves superimposed on the ground swell,

TABLE 3.4

STATE-BY-STATE EFFECT OF THE RELIGIOUS ISSUE
(DEMOCRATIC RESULT AS PERCENTAGE OF TOTAL VOTE)

State	Electoral Vote	Simulation Prediction with No Protestant Shift and No Catholic Shift	Anti-Catholic Shift by Protestants	Catholic Shift	Best-Fit Simulation Prediction	Actual Result	Net Effect of the Religious Issue
Oklahoma	8	.764	−.200	.007	.572	.410	−.192
North Carolina	14	.740	−.196	.012	.556	.521	−.184
Arkansas	8	.746	−.194	.012	.563	.539	−.182
Georgia	12	.728	−.191	.015	.552	.626	−.176
Virginia	12	.736	−.189	.015	.562	.473	−.174
South Carolina	8	.720	−.185	.018	.554	.512	−.167
Alabama	11	.725	−.184	.018	.559	.577	−.166
Florida	10	.731	−.182	.019	.568	.485	−.163
Mississippi	8	.708	−.176	.024	.556	.596	−.152
Texas	24	.747	−.161	.024	.610	.510	−.138
West Virginia	8	.573	−.080	.016	.509	.526	−.064

(*Continued*)

119

TABLE 3.4 (Continued)

State	Electoral Vote	Simulation Prediction with No Protestant Shift and No Catholic Shift	Anti-Catholic Shift by Protestants	Catholic Shift	Best-Fit Simulation Prediction	Actual Result	Net Effect of the Religious Issue
Kentucky	10	.578	− .076	.022	.524	.464	− .054
Tennessee	11	.557	− .083	.030	.503	.464	− .053
Louisiana	10	.714	− .100	.048	.662	.638	− .052
Utah	4	.473	− .060	.009	.421	.452	− .052
Idaho	4	.458	− .061	.011	.408	.464	− .050
Oregon	6	.483	− .054	.018	.447	.475	− .036
Kansas	8	.451	− .057	.023	.417	.393	− .034
Indiana	13	.459	− .057	.024	.426	.448	− .033
Washington	9	.491	− .053	.020	.459	.491	− .032
Iowa	10	.451	− .055	.023	.419	.431	− .031
Maryland	9	.577	− .072	.042	.547	.538	− .030
Wyoming	3	.477	− .054	.025	.448	.449	− .028
Nebraska	6	.458	− .054	.027	.431	.385	− .027
Nevada	3	.480	− .053	.029	.456	.512	− .024
Delaware	3	.593	− .065	.042	.570	.508	− .023
South Dakota	4	.448	− .053	.030	.425	.417	− .023

Missouri	13	.477	−.055	.033	.455	.507	−.022
Colorado	6	.503	−.048	.029	.485	.458	−.019
Ohio	25	.486	−.051	.035	.470	.468	−.016
Montana	4	.487	−.049	.034	.472	.487	−.015
North Dakota	4	.454	−.050	.036	.440	.446	−.015
Minnesota	11	.479	−.047	.036	.468	.507	−.012
Michigan	20	.497	−.049	.042	.490	.511	−.007
California	32	.526	−.043	.038	.522	.497	−.005
Arizona	4	.508	−.042	.043	.509	.445	.001
Wisconsin	12	.490	−.043	.044	.492	.481	.002
Illinois	27	.510	−.045	.049	.515	.501	.004
Maine	5	.372	−.019	.041	.394	.430	.022
New Mexico	4	.514	−.035	.061	.539	.502	.026
Pennsylvania	32	.434	−.021	.051	.464	.513	.030
New York	45	.500	−.026	.064	.538	.528	.038
New Jersey	16	.478	−.022	.062	.518	.504	.040
Vermont	3	.384	−.011	.052	.425	.414	.040
New Hampshire	4	.406	−.014	.056	.448	.466	.042
Connecticut	8	.480	−.015	.072	.536	.537	.056
Massachusetts	16	.495	−.015	.077	.557	.605	.061
Rhode Island	4	.516	−.012	.088	.592	.638	.076

which was the religious issue. Specifically, in the South the defection from Kennedy by Democrats was larger than could be explained by their attitudes on the religious issue alone.

A rational model of this defection in the South would have to incorporate the fact that not everyone in the South was reacting to the Negro awakening in the same way. Specifically, Negroes were moving from their traditional Republicanism into the Democratic camp as they had been doing increasingly ever since the Roosevelt era. At the same time and for the same reasons, white Democrats were moving out of it. So we added to our model an additional variable, applied in the South only, designed to represent the complex of civil rights and race issues.

We postulated that some of the Southern whites whom the rest of the simulation equations would have had vote for Kennedy would shift to Nixon and some of the nonwhites whom the rest of the simulation equations would have had vote for Nixon would shift to Kennedy. Empirically we find that a value for this shift parameter of about 10 per cent pulls the simulation of the South into line. That is to say, if after party tradition and the religious issues had done their work, one allows 10 per cent of Southern voters to shift in appropriate directions, with the few Negro voters partly offsetting the white shift, one gets a good prediction.[6] That represents the added extent to which the Southern rebellion against Northern liberal Democracy expressed itself by a bolt of some voters to the Republicans.

[6] An estimated 1,414,000 Negroes were registered to vote in the South in 1960 compared to only about 1,000,000 in 1952. The 1960 figure represents about 28 per cent of Southern Negroes of voting age. Harold F. Gosnell and Robert E. Martin, "The Negro as Voter and Officeholder," *Journal of Negro Education* (Fall 1963), pp. 415–425.

5. Residual Variations

The special Southern vote adjustment improved the results of the simulation, but only modestly. The co-efficient of correlation of the simulation and election results state by state for the 48 states, the South included, remained only .70. The coefficient of correlation for the 11 Southern states alone was only .28. In short, while the correction pulled the region as a whole into line it did not order the individual states well. This was no surprise. As we had said in August 1960, "so much in the South depends on what a few leaders do that a meaningful *state-by-state* simulation would have to include these variables." And that was not possible from the data base we were using. We had no information about state political leadership or maneuvers.

But there is still something to be learned by noting what states did not fit the model well. That helps identify the places where local events of significance were taking place. There are various reasons why a single state may deviate more than most. It may be random variation; in any distribution some items are bound to be bigger than others. It may be due to errors in the data base; an error could occur in the original poll cards we used, in our recoding of them, in the census figures about the state as we transcribed or manipulated them, in machine output, or in any one of a large number of other places. Or finally, deviation of the vote in any state from prediction may be due to true political idiosyncrasies; and that is what interests us.

The most notable idiosyncrasy is to be found in three states that are the core of the deep South: Georgia, Alabama, and Mississippi. These are states where unreconstructed segregationism is most powerful. Yet these three states gave Kennedy a margin far greater than did the rest of the South.

No single simple social hypothesis accounts for this result. The unreconstructed deepest South also includes South Carolina, which despite its Governor's strong support of Kennedy failed to vote for the ticket any more strongly than did such less deeply Southern states as North Carolina or Virginia. The three strongly pro-Kennedy states include one with but limited industry, Mississippi, but two where the plantation rural areas surround a major industrial center with a large laboring class, Atlanta, Georgia, and Birmingham, Alabama. They include two states which send many archconservatives to Congress, Georgia and Mississippi, but one, Alabama, which has been represented by such liberals as Senators John Sparkman and Lister Hill.

Thus, no single simple social characteristic divides the South in the same way that it divided politically into two sets of states. We are forced to recognize the significance of local leadership, tradition, and organization.

There is, however, a more general observation that can be made. Despite the absence of South Carolina from this group of strongly Democratic states, one cannot help but be struck by the fact that this group does include three of the four hard-core Southern states. Kennedy did better in this heartland of the South than in those areas where its traditions have been more rapidly wearing away.

This suggests an interesting phenomenon. The more deeply Southern an area, the more Democratic it is and at the same time the more antiliberal and anti-Catholic. In short, the more Southern it is the more deeply it is torn between dislike of what a Northern liberal Democratic candidate stands for and distaste at voting Republican. In the deep South both these feelings are more intense and so is the tension between them. But which of the two will win out is indeterminate. There is no a priori conclusion as to whether the deepest South areas will bolt more than the more moderate areas against a pro-civil rights Democratic candidate, or less. We have already suggested that the answer will depend in part on organization. The Governor and party leaders may push a deep South state either way. Choosing uncommitted electors or voting for a third, Dixiecrat, party helps resolve the tension between pressures. But on balance, when such options have not been available, it seems clear that being part of the deep South has exerted more pressure toward Democratic party loyalty than toward revolt. The most segregationist Democrats are most captive of their one-party tradition.

This conclusion is supported by the surprising Kennedy vote in Georgia, Alabama, and Mississippi. It is also evidenced by some figures from our data bank which show that the more closely a person is embedded in the traditional community organization of the South the more likely he is to vote Democratic. (See Table 3.5) Men (who are more involved in the political-economic management of civic life in Southern communities than women) were more likely than were women to say they would vote for a Democrat against Nixon and a Democrat for Congress. Small-town and rural people who were more part of the traditional Southern system were also more likely than were dwellers in the anomic big cities to say they would vote for a Democrat against Nixon and for a Democrat for Congress. And so, at the extremes, it was big-city women who were most likely to break with the South-

TABLE 3.5

LOCALE OF EMERGENCE OF A TWO-PARTY
SYSTEM IN THE SOUTH

Percentage* Willing to Vote for:	Rural		Town		Metropolitan	
	Men	Women	Men	Women	Men	Women
Nixon	31	37	32	32	38	47
A Republican for Congress	20	23	20	23	21	26

* Excluding undecided. Complement is percentage unwilling.

ern tradition and vote for Nixon, and it was small-town and rural men who were most solidly Democratic.

Politically astute readers will undoubtedly spot reasons for some of the larger deviations from the model found in other states (see Table 3.1). Massachusetts, for example, had a personal attachment to the late President and to his family. But we shall not try here to rationalize each deviation. That would require institutional and historical data not part of our particular archive. Our purpose in discussing the deviant cases is to show by example how this may be done.

Our broader purpose concerns not so much the deviant cases but rather the total fit. Hopefully, the reader now sees how postdiction with a model the parameters of which are chosen with a view to a good fit can help us understand what is going on in the real world. Our computer simulation model, while complex compared to most verbal models, is simple compared to the real world. It is thus a fruitful compromise, complex enough to account with some accuracy for most of the variance in the real world, but simple enough to permit rational understanding by our limited human intellects.

4

Do Predispositional Factors Summate?

Did we need as complex a model as we have been
using to analyze the behavior of the electorate in 1960?
Could the same results have been reached with a far
simpler set of figures than the 360,000 in our data bank?

Simulation models are appropriately subject to such
a challenge. Computer simulations are generally so
complex that it is impossible to analyze them fully.
More parsimonious models are to be preferred if they
can do the job. In our study in 1960 we had no simple
theory for predicting how each of the 480 voter-types
would feel about each of the 50 issues, so we compiled
data on $480 \times 50 \times 3 = 72,000$ empirical variables. In
the simulation we actually conducted during the campaign we looked up and entered values for about 1,500
of these from the data bank.

Clearly a model that could start with fewer empirically counted numbers and predict the rest of the
numbers on some statistical basis would be more ele-
gant than the brute force method we used. It is therefore appropriate, in retrospect, to see if such a basis
can be found.

Our vast data matrix permits us to attack a problem not previously feasible to investigate: the question of the *additivity* of predispositional factors. Catholics are typically more heavily Democratic than Protestants; city dwellers more heavily Democratic than country dwellers; low socioeconomic groups more so than high socioeconomic groups. Does this mean that Catholic urban workers are triply more Democratically inclined? Or is there perhaps an intensifying or multiplicative relationship so that when three factors impel in the same direction then there is even greater preference for the favored party than one would expect from the three factors acting individually? Some years ago, Lazarsfeld and Berelson[1] devised an "Index of Political Predisposition" for rough-and-ready purposes of survey analysis. This Index assigned two points for being Catholic, one point for being urban, and so on, and considered total pro-Democratic predisposition to be the sum of these points. This seems a plausible albeit crude thing to do, but no one has ever precisely investigated the presumption upon which this device is based, namely that the predisposing factors summate.

Analyses of variance techniques are useful for analyzing problems of this kind for they allow the effects of several different independent variables on a single dependent variable to be separated from one another. They allow us to separate the effects of the independent variables (sex, religion, region, socioeconomic status, community size, and race) on vote or on any other dependent variable. For example, more Catholics vote Democratic than Republican. Should one assume from this fact that the "Catholic" effect on behavior is to increase Democratic disposition? Not necessarily, for

[1] Paul F. Lazarsfeld, Bernard Berelson, and Hazel Gaudet, *The People's Choice* (New York: Columbia University Press, 1948).

at *a given income level or in a given region* Catholic disposition for the Democratic party may not necessarily be any greater than for Protestants. The percentage of Protestants and Catholics belonging to the Democratic party could be the same at every income level; the greater percentage of Democrats among Catholics could then be attributed to the lower-income distribution among Catholics.

The result of an analysis of variance which sorts out the impact of each single factor separately is directly relevant to sociological theories of voting, but, as already suggested, is only marginally relevant to election predictions.

In discussing the effect of independent variables it is necessary to introduce the somewhat subtle notion of interaction. Two or more independent variables may affect a dependent variable in ways that are simply additive, or they may work together in a way that mere summation would not suggest; that is, there may be an interaction.

The idea of a statistical interaction in the data can be explained by reference to an imaginary experiment on the effect of cream and sugar on coffee consumption. Suppose there is a group of ten people to whom on four days free coffee is offered; the first day neither cream nor sugar is present, the second day cream is present, the third day sugar is present, and the fourth day, the "cream-and-sugar day," both cream and sugar are present. On the first day, when they have to drink the free coffee black, perhaps only two members of the group, Able and Baker, take a cup. We would probably find that on the second "cream, no-sugar" day coffee consumption was higher than on the "coffee-only" day. Perhaps Able, Baker, and Cantor take coffee. On the third or "sugar, no-cream" day coffee consumption would also be higher than on the "coffee-only" day. Perhaps Able, Baker, and Ewing take a cup. We would probably also find that on the day that both cream and sugar were available, coffee consumption would be significantly higher than on the other days and higher than just the sum of the coffee drinkers under the separate black, cream, and sugar conditions. Able, Baker, Cantor, and Ewing could all now have coffee as they liked it, but so could perhaps nine of our ten men (the tenth just doesn't want coffee). So a cream condition produces three drinkers, a sugar condition three also, but a sugar-and-cream condition produces nine, not six. This is an example of interaction. The effect of the cream on coffee consumption and the effect of sugar on coffee consumption do not completely explain the effect on coffee consumption of "cream and sugar," that is, the effects are not additive. In this case we would say that cream and sugar together had an effect on coffee consumption that was different from the sum of their individual effects.

When analysis of variance reveals that the effects of the independent variables summate, it is possible to reconstruct the dependent variable on each given voter-type from linear arrays of the independent variable. However, if there are interactions among the variables, two-, three-, or four-dimensional arrays must be referred to in reconstructing from marginals the dependent variable for a voter-type.

In studying the distribution of attitudes, or the distribution of attitude changes, among the voter-types, we are attempting to see if one can reconstruct an array of five dimensions—region, socioeconomic status, religion, sex, and party, from arrays of fewer dimensions. Of course, there may be no number of arrays with less than five dimensions from which the array can be reconstructed. At the other extreme, it may be possible to reconstruct the original array from five or less arrays of one dimension, that is, linear arrays. If that is possible then we may say that factors summate.

For the present analysis, a five-way factorial array was constructed, using 336 of the 480 voter-types and eliminating those creating difficulties in forming a neat array. Table 4.1 displays the five factors: party identi-

TABLE 4.1

FACTORS USED IN THE ANALYSIS

Party identification	Democratic, Republican, Independent
Religion	Catholic, Protestant
Region	East, South, Midwest, West
SES	Rural A or B, Rural C
	Town A or B, Town C,
	Urban A, Urban B, Urban C
Sex	Male, Female

Note: A means professional, executive, and managerial occupations; B means white-collar occupations, C means blue-collar occupations.

Urban refers to metropolitan areas exceeding 100,000 population; town refers to communities between 5,000 and 100,000; rural refers to places under 5,000.

fication, religion, region, sex, and socioeconomic status, the latter referring to our combination of occupation and city size.

When all combinations of these five factors are formed, we obtain a $3 \times 2 \times 4 \times 7 \times 2$ factorial design with 336 cells. In each cell is entered a score on an issue-cluster for individuals of that cell type. Of the 52 issue-clusters we here analyze the 32 with sufficient data for the five-way analysis, breaking down as follows: attitudes toward political figures (Truman, McCarthy, and so on), 5 clusters; trial heats between potential candidates (for example Kennedy versus Nixon), 2 clusters; Presidential and Congressional voting records, 5; "images" of the parties (for example, "Which party is more likely to keep America out of war?"), 4; foreign-policy attitudes (United Nations, foreign aid, summitry, and so on), 9; domestic issues (labor policy, civil rights, and so on), 3; involvement in elections, 3; and finally a cluster on anti-Catholic sentiment.

Each of these issues allows two definite sides or positions and a "don't know" or "undecided" response. To get a single score for a cell for each cluster, the proportion of its members favoring the Republican side of the issue was subtracted from the proportion favoring the Democratic side. Thus the dependent variable is the net proportion, $P - Q$, of the cell members favoring the Democratic side, omitting the Undecideds. This net proportion is "plus" or "minus" depending on which way is "up" for the issue, and the following arbitrary convention was adopted: "up" is the liberal direction.

Having computed net proportions for each of the 336 cells on a given cluster, each array was then subjected to a 5-way analysis of variance. This yielded 5 main effects, 10 first-order interactions, 10 second-order interactions, 5 third-order interactions, and 1 fourth-order interaction for each issue-cluster. Our interest is in whether the interactions, particularly the more readily conceptuable first-order interactions, do or do not generally run higher than one would expect on the basis of sheer sampling error in the original net proportions. Before looking at this, however, let us backtrack a bit for clarification. We want to be sure that the notion of additivity of factors (that is, absence of significant interactions) is well understood. Table 4.2 concerns attitudes toward Stevenson. The dependent variable is the proportion giving favorable responses

TABLE 4.2

ATTITUDE TOWARD STEVENSON

$P - Q$: Mean of all cells: .087

Main effects:

Party: Democrat, .455; Republican, −.432; Independent, −.023
Region: East, .024; South, −.044; Midwest, −.003; West, .023
SES: Rural A, B, −.044; Rural C, −.032;
          Town A, B, −.009; Town C, .016
          Urban A, .007; Urban B, .000; Urban C, .062
Religion: Protestant, −.022; Catholic, .022

Demonstration of additivity:

The combination Republican Midwestern Town A + B Protestant is additively constructed as .087 − .432 − .003 − .009 − .022 = −.379
Actual = −.3581

---

toward Stevenson minus the proportion giving unfavorable responses. The table displays the effects found to matter in the analysis of variance, namely party, geographic region, occupational and city-size level, and religion (sex yielded no differences, and all interactions turned out small and not significant). The mean of all 336 cells is .087 (which incidentally is biased upward because on poll questions requiring like-dislike responses, many respondents avoid the "dislike" response). The effects of party, averaged over 112 cells for each of the three parties, are given in the next row as deviation scores. The next row gives regional effects, the third row gives the level effects, and the fourth row the religious effect. The last two rows illustrate that additivity does obtain for a particular voter-type on this cluster, that is, the interactions are not significant. To reconstruct the net proportion in a subgroup, for example, Republican Midwestern Town A + B Protestant one simply adds the over-all mean to the respective main effects. This illustrative calculation yields −.379. The actual net proportion in that group was —.3581 and the discrepancy is well within the limits of sampling error. All other net proportions can be similarly reconstructed. Where additivity prevails the main effects provide a very neat, parsimonious description of the results for all combinations.

Picking up the main thread again, and getting into the unavoidable technical details, we wish to know whether the interactions for other clusters can be so deftly disposed of as with the example just given. The analysis of each cluster yields many different interactions of various orders. The 10 first-order interactions are the most important. Statistical significance tests for these interactions could in principle be devised with a theoretical error term dependent only upon sample sizes and binomial assumptions. However, this is rather tricky to do, for reasons that will not be broached here. The simpler alternative is to pool second-, third-, and fourth-order interactions to produce an error mean square for each cluster. This is what was done. Each of the 5 main effects and 10 first-order interactions for each cluster were then F-tested against the single error term for the cluster. The result of this massive statistical assault was to yield 32 F-ratios for each main effect and each first-order interaction.

Table 4.3 gives some indication of the size of these F-ratios. Each entry in the table gives the *median* F-ratio (over the 32 issue-clusters) for a particular effect. Notice that the median F-ratios are high for the main effects, and are uniformly unimposing for all interactions, hovering somewhere around 1.00. Actually this display, though aesthetic, is somewhat misleading, since the theoretical median of an F-distribution is somewhat below 1.00, by an amount depending on degrees of freedom.

A more searching examination of the first-order interactions may be made by using the entire distribution of F's rather than just the median. One may take each *obtained* distribution of 32 F-ratios and compare it with the theoretical distribution drawn from extended versions of the usual F-table. The purpose of this comparison is as follows: If there is really no general

TABLE 4.3
MEDIAN F-RATIOS FOR INTERACTIONS
(MAIN EFFECTS ON DIAGONAL)

	Party	Religion	Region	SES	Sex
Party	(36.09)	1.32	1.29	1.09	1.06
Religion	1.32	(4.78)	1.08	1.05	.52
Region	1.29	1.08	(4.02)	1.15	.88
SES	1.09	1.05	1.15	(3.67)	.86
Sex	1.06	.52	.88	.86	(1.88)

tendency toward systematic interactions, then the distribution of obtained F-ratios will approximate the known form attributable to sampling errors. If, on the other hand, there is a general systematic tendency toward interactions, the obtained distribution of F-ratios will deviate from the known form in the direction of showing too many large F-ratios.

Before the comparison is made, it is useful to distinguish two types of issue-clusters: those for which the main effect of party identification comes out impressively large, and those for which it does not. In the former category, which we shall call party-dominated issues, fall voting responses, party images, attitudes toward political figures, and most domestic issues; in the latter category fall the foreign-policy issues, civil rights, and anti-Catholic sentiment.

Table 4.4 shows part of the comparison of observed

F's with theoretical F's. Here are tabulated the results for 4 of the 10 first-order interactions: party $\times$ religion, party $\times$ region, party $\times$ SES, and party $\times$ sex. These are pooled into a single comparison. For party-dominated issues, it is quite clear without belaboring it with a chi-square test that too many F-ratios fall in the

TABLE 4.4

INTERACTIONS OF PARTY WITH OTHER FACTORS
(F-DISTRIBUTIONS: OBSERVED VERSUS EXPECTED)

	Low 10 Per Cent	Next 15 Per Cent	Next 25 Per Cent	Next 25 Per Cent	Next 15 Per Cent	Top 10 Per Cent
Party-Dominated Issues						
Observed	4	1	8	14	10	27
Expected	6.4	9.6	16.0	16.0	9.6	6.4
Other Issues						
Observed	8	11	8	19	9	9
Expected	6.4	9.6	16.0	16.0	9.6	6.4

upper portion of the theoretical distribution. In other words, for *party-dominated issues, there is a strong tendency for party identification to interact systematically with other background factors.* The neat additive property shown in Table 4.2 cannot be said to hold generally.

However, at the bottom of the table we see the results for non-party-dominated issues. Here there is a fairly good correspondence between observed and expected values. Deviations of observed from expected are not statistically significant, and it can be concluded that for *non-party-dominated issues there is no discernible general tendency for party to interact with other background factors.*

In Table 4.5 are displayed the results for the 6 first-order interactions not involving party, for example, religion × region, and so on. Summarizing this briefly:

Neither for party-dominated or non-party-dominated issues is there a generally discernible tendency for two factors other than party to interact. This statement is statistically borderline enough to test it by chi-squares,

TABLE 4.5
INTERACTIONS OF FACTORS OTHER THAN PARTY
(F-DISTRIBUTIONS: OBSERVED VERSUS EXPECTED)

	Low 10 Per Cent	Next 15 Per Cent	Next 25 Per Cent	Next 25 Per Cent	Next 15 Per Cent	Top 10 Per Cent
Party-Dominated Issues						
Observed	9	13	25	22	10	17
Expected	9.6	14.4	24.0	24.0	14.4	9.6
Other Issues						
Observed	7	12	20	28	16	13
Expected	9.6	14.4	24.0	24.0	14.4	9.6

inasmuch as there is some suggestion of a slight excess of large F's. Using a very sensitive test[2] that takes into account the pileup of cases toward the tail of the distribution, the chi-square was 1.29 for party-dominated issues and 2.99 for non-party-dominated issues, both values nonsignificant on 1 degree of freedom. It is thus generous to concede a violation of the null hypothesis on the basis of the data here. By conventional standards one would say that even for non-party-dominated issues, the frequency of large F's is not excessive and that one cannot assert that there is any general mechanism tending to produce systematic interactions

[2] John W. Tukey, "Some Possible General Idiosyncrasies in the Use of Statistical Techniques and Quantitative Approaches," Mimeographed paper, pp. 106–109.

across all issues and factors. Statistics aside, at the level of specific issue-clusters and factor combinations, one may identify occasional interactions that are undoubtedly true effects, for example, the interaction of region and religion on the civil-rights cluster. (Southern Catholics are nowhere near as anti-integration as are Southern Protestants.) However, such examples are quite exceptional.

Thus we may respond to the rhetorical question in the chapter title, "Do predispositional factors summate?" with the answer, "By and large, yes." However, for party-dominated issues, party identification quite often interacts with other factors; that is, party identification does not always summate with other factors in determining issue response.

What is the character of the systematic interactions of party with other factors on party-dominated issues? There seem to be two types of such interactions: one, a shrinkage effect applied to party polarity within certain social types, and the second, a shift effect applicable to Independents of certain social types.

Table 4.6 displays two-way arrays for which close inspection reveals the shrinkage effects on the Attitude Toward Truman cluster. The effects on this cluster are reasonably representative of this phenomenon as it appears on other clusters as well. What happens is that party differences in attitudes toward Truman are less sharp in the South than elsewhere, are less sharp among C occupation groups than among A and B groups, are less sharp among Catholics than among Protestants, and among women than among men. In short, all the social groupings which typically display less involvement and interest in elections also display less polarization of opinions around representative political objects.

The second common type of party interaction is

TABLE 4.6

ILLUSTRATIVE INTERACTIONS OF PARTY WITH OTHER FACTORS:
ATTITUDE TOWARD TRUMAN CLUSTER

	Democrat	Republican	Difference: Democrat–Republican
East	.518	−.531	1.049
South	.316	−.476	.792
Midwest	.534	−.543	1.077
West	.470	−.548	1.018
A + B Rural	.447	−.579	1.026
C Rural	.455	−.367	.822
A + B Town	.418	−.621	1.039
C Town	.494	−.474	.968
A Metropolitan	.418	−.637	1.055
B Metropolitan	.481	−.613	1.094
C Metropolitan	.505	−.385	.890
Protestant	.439	−.597	1.036
Catholic	.480	−.452	.932
Male	.502	−.553	1.055
Female	.417	−.497	.914

Note: Entries in the first two columns each give average $(P - Q)$ values for all voter-types of the given classification.

TABLE 4.7

ILLUSTRATIVE INTERACTIONS OF PARTY WITH OTHER FACTORS:
CONGRESSIONAL VOTE-CLUSTER

	Democrat	Republican	Independent
East	.771	−.830	−.087
South	.838	−.632	.199
Midwest	.819	−.879	−.022
West	.753	−.838	−.070
Protestant	.792	−.843	−.061
Catholic	.799	−.746	.071

Note: Entries give average $(P - Q)$ values for all voter-types of the given classification.

illustrated in Table 4.7 for the Congressional Voting cluster. The effect is that Independents behave like Democrats among social types with strong Democratic traditions (Southerners, Catholics), but that otherwise they lean very slightly toward the Republicans.

Such party interactions as we have indicated, while clear, plausible, and statistically significant, are small in magnitude compared to the main effects, that is, the effects due to the single classifications alone. Having performed the major analysis described in this chapter, it is worthwhile to set forth which main effects make a sizable difference in which clusters. In Table 4.8, all main effects significant at the .01 probability level are indicated by a tally opposite the appropriate cluster. In general there are few surprises in Table 4.8, and we leave the bulk of the digestion of the table to the reader. We note in passing, however, the foreign-policy phenomenon referred to in Chapter 3, namely that foreign-policy opinions cross-cut party affiliations and frequently lack substantial correlation with social characteristics as well. This may simply be a reflection of the general lack of commitment of the public in foreign policy, particularly on bland issues which do not engage any heat except at the fringes of the ideological spectrum. Note especially the lack of any significant main effects at all for Cluster 41, Meet with Soviets, and Cluster 49, Attitude Toward U.N.

Finally, let us return to the question of analytical parsimony raised at the beginning of the chapter. Could we have made economies in the use of the data bank by capitalizing on any possible additivity of factors for the clusters used? We must ask two questions: First of all, did additivity obtain for any of the clusters used in any of our simulations, and if it did, in what way could it have been of use to us?

TABLE 4.8
Significant Main Effects for Thirty-Two Clusters

Cluster Number	Region	SES	Religion	Sex	Party
1. Attitude Toward Truman	X*	X	X	0†	X
2. Anti-McCarthyism	0	X	X	0	X
4. Anti-Nixon Voting	0	0	X	0	X
5. Attitude Toward Eisenhower	X	X	0	X	X
6. Attitude Toward Stevenson	0	0	0	0	X
8. Kennedy versus Nixon	0	0	X	0	X
9. Pro-Negotiationism	X	0	0	0	0
14. Anti-Catholic	0	0	X	0	X
16. New Deal Philosophy	0	0	0	0	X
17. Civil Rights	X	X	X	0	X
19. Congressional Votes	X	X	X	0	X
20. Which Party for Prosperity	X	X	X	0	X
21. Which Party for Personal Finances	0	X	X	0	X
23. Best Party for People Like Self	0	X	X	0	X
24. Normal Presidential Vote	X	X	X	0	X
26. Stevenson versus Eisenhower (1952)	X	X	X	0	X
27. Stevenson versus Eisenhower (1956)	X	X	0	0	X
29. Fear of War	0	X	0	0	0
31. Concern over Wages, Unemployment	0	0	0	0	X
32. Foreign Policy Knowledge	X	X	0	X	0
34. Interest in Elections	X	X	0	X	X
35. Non-Voting	X	X	0	X	X
37. Voting Intent Record	X	X	0	X	X
38. Ike Vote but Democrat for Congress	X	0	0	0	X
39. Which Party Best for Peace	0	0	0	0	X
40. World Situation Complacency	X	0	0	0	0
41. Meet with Soviets	0	0	0	0	0
42. Red China Policy	0	X	X	X	0
46. Defense Commitments	X	0	0	X	0
49. Attitude Toward U.N.	0	0	0	0	0
50. Attitude Toward Foreign Aid	X	X	0	0	X
51. Nixon versus Rockefeller	0	0	0	0	X

* An X indicates a significant effect
† An 0 indicates a nonsignificant effect.

The answer to the first question is that on the anti-Catholicism cluster and the negotiationism cluster, there is almost perfect additivity of factors, but on the Congressional and Presidential voting clusters and the turnout cluster, there is (as we have indicated) some lack of strict additivity. The second question merits a brief discussion.

In general, the equations of our simulations involved *products* of voting-preference statistics for a voter-type and issue-impact statistics (and turnout statistics as well). Thus even if strict additivity obtained for all clusters involved in the equations, it would still have been necessary to perform a separate calculation for each of the 480 voter-types, rather than fewer calculations on aggregated types. Products of additive statistics are no longer additive (and there is the additional complication of differential weighting of different voter-types). Thus we were on safe ground in putting the computer to the trouble of aggregating results over all 480 types. However, there is another way in which we could have made use of the fact of additivity for the anti-Catholicism cluster had we been cognizant of this fact at the time of the simulation.

The Catholic issue simulation required estimates of P_{14}, the proportion unwilling to vote for a Catholic in each of 285 non-Catholic voter-types. Many of these 285 estimates were based upon a small number of individuals, and thus were not very stable statistically. The statistical stability could have been improved by using the fact of additivity, which enables the "borrowing" of data from types similar to a given type. That is, the entire array of P_{14}'s could have been estimated by summating the main effects of the separate factors defining the type, in the fashion illustrated in Table 4.2. This would have required the use of only 13 independent main-effect numbers to estimate P_{14} for 168 of the 285 non-Catholic types. The remainder of the types do not fall into a very neat array (nor do we know for sure that additivity of anti-Catholic tendencies holds among Jews, Negroes, and so on), and they would have to have been handled individually as before, or by special devices. In any case, considerable statistical efficiency could have been achieved by using the additivity property to estimate the P_{14}'s. (Following this estimation step, of course, they would enter the simulation equations for each of the 480 voter-types separately as before.)

It seems likely that good future analytical use can be made of the tendency discovered in our analysis for social factors to be substantially additive for large portions of public opinion data.

5

Political Trends of the 1950's

INTRODUCTION

During the Eisenhower Administration, the first Republican administration since 1932, American life was changing. Did political alignments change also? Our data bank with its five biennial matrices and its 130,000 cases lent itself to exploring what changes if any were taking place in the political characteristics of the public.

We singled out 5 of the 50 issue-clusters on which to study time-series trends in public opinion. These 5 seemed likely to shed light on any political changes taking place in the 1950's. The issue-clusters analyzed were: evaluation of Eisenhower's performance as President; attitude towards Stevenson; vote for Congress; candidate preference in the 1952 election; and candidate preference in the 1956 election.

The trends described here are attitude changes over time for 336 of the voter-types. All voter-types in the border states and some smaller voter-types in other regions (such as Negroes and Jews) were omitted from the analysis (as in the previous chapter) for reasons of statistical convenience. In this analysis of trends no weighting is given for size of the voter-types. When a statement about trend among Democrats is made the statement refers to the average trend among the 112 Democratic voter-types included in this portion of the study. The trend must apply to all kinds of Democrats, not just to a few Democratic voter-types (such as Southern Democrats or urban lower-income Democrats) who by virtue of their numbers cast their stamp on the Democratic total. If our purpose here were prediction, as in the 1960 simulation, weighting and accuracy proportional to the size of the group would be necessary since large groups affect the outcome of elections more than do small. Here, however, we are interested in generalizing about groups; small subsets of the groups can provide just as critical observations as can large ones. We wish our statements about trend to apply uniformly at least to specified subsets of the voter-types.

In general, we found in the previous chapter that with limited exceptions public opinion data do tend to summate: the variables which define the voter-types have independent effects on the attitudes of the voter-types, and these effects, taken one at a time largely account for the joint effects. Once this is established, it is possible to compare issue-clusters to see the relative importance of different factors in accounting for the distribution of attitudes on different types of issues. For example, the religious factor did not strongly affect attitudes toward Stevenson. In contrast, there was a strong religious factor in attitude towards Truman. When we understand on what types of issues a factor —the religious factor—for example, is most important, or is unimportant, we have added to sociological knowledge. In the present chapter, we attempt to apply this technique to obtain knowledge of trends over the period 1950–1960.

DEPENDENT VARIABLES

As mentioned previously, all issue-clusters were trichotomized into the percentage of respondents giving Democratic responses to the survey questions, called P, the percentage giving Republican answers, called Q, and the residual category, the percentage giving "don't know" or obscure answers. For any issue-cluster two dependent variables could be used to study trends: $(P - Q)$, or $(P + Q)$. The first variable, $(P - Q)$, which may take all values from $+1$ to -1, may be considered a measure of attraction to the position of the two parties. If 100 per cent of a group take the Democratic position on an issue, then $(P - Q)$ assumes its maximum value, $+1$, and similarly, when 100 per cent of a voter-type take the Republican stand on an attitude, the variable takes on its lowest value, -1. The second dependent variable, $(P + Q)$, which may take all values between $+1$ and 0, can be considered a measure of commitment to a position. For example, if every survey respondent of a given voter-type gave "don't know" responses to questions composing an issue-cluster, $(P + Q)$ would assume its lowest value, 0, while if every respondent of a given type took a stand on the issue, $(P + Q)$ would assume its maximum value, $+1$.

DIFFERENTIAL AND ACTUAL TRENDS

There are two types of trends discussed in this chapter: differential trends between groups and actual trends. Differential trends are measured by divergence of the group trend from the mean trend for all 336 voter-types. By limiting ourselves mainly to differential trends, the artifact of variations from year to year in question wording is made an unimportant problem for us because the dependent variable used takes off from the different question levels in different years. For example, suppose the over-all average for a variable in one year is .8 and in another year .6. If one voter-type has a level of .7 in both years that group has, in an absolute sense, stood still. In the sense of our study, however, the group has moved differentially in comparison with the mean. Before drawing the conclusion that they stood still and that the rest of the population moved, we should recall that our measuring instrument is not stable. Affirmative answers to questions in the first year at the level of .8 do not necessarily mean something different than .6 affirmative answers to other questions in a second year. (From outside the solar system it is all relative whether we say the earth moves around the sun or the sun moves relative to the earth.)

TRENDS IN CONGRESSIONAL VOTING

The survey data that we are using permit comparisons of Congressional voting behavior that cannot be made from election statistics. The official records do not distinguish who cast the votes except by geography. Surveys permit us to distinguish, for example, how Democratic, Republican, and Independent voters differed in the votes they cast for the period covered by the data bank.[1] In a country where the majority of

[1] Tabulation of Congressional votes for the 43 states studied here, the border states having been excluded, show that the Democrats received 49.9 per cent of the total Congressional vote in 1952, 50.8

voters have basically Democratic leanings, General Eisenhower was able to win two convincing victories, achieving strong support in the South and among Independent and Democratic voters. In 1952 he carried the House with him, but in 1956, even with Eisenhower virtually sweeping the country again, the Republican party was unable to carry either house of Congress, the first time this has ever happened. The data presented here suggest that GOP Congressional fortunes did not fall because of any change in voting habits by Republicans, but because many Democrats and Independents who had voted for GOP Congressmen in 1952 steadily shifted their preferences to Democrats thereafter and also began to vote more frequently than previously.

Not only are the majority of Americans Democrats, but also a slight majority of Independents usually vote Democratic. In 1952 and 1954, however, the Independent vote favored the Republican party.

Between the 1951–1952 surveys and the 1959 surveys the percentage of Democratic and Independent voters who had a preference for either party in Congressional elections rose steadily. At the same time, the percentage of those with a choice who preferred the Democratic party also rose. Over this time period, there was no over-all shift in Congressional vote preference among Republicans. Analysis of Democrat and Independent voter-types, however, shows average decreases of .02 and .025 per year, respectively, in Republican preferences (Q), and average increases of .038 and .04 per year in Democratic preferences. Not only did Democrats tend toward more sharply focused party lines and

---

per cent in 1954, 50.1 per cent in 1956, 55.9 per cent in 1958, and 54.6 per cent of the total Congressional vote in 1960. Such over-all trends are better documented by election statistics than by the survey data we are using here.

TABLE 5.1
1952 CONGRESSIONAL PREFERENCE

Party	Prefer Democrats (P)	Prefer Republicans (Q)	(P + Q)	(P − Q)
Democratic	.77	.10	.87	.67
Independent	.29	.39	.68	−.10
Republican	.07	.81	.88	−.74
Average	.38	.43	.81	−.05

Independents move from favoring Republicans to favoring Democrats, both also tended toward an increase in total voting intention (as measured by the sum $P + Q$). For example, the average of $(P + Q)$ for the 112 Independent voter-types was .68 in 1952 and .80 in 1956. (See Tables 5.1 and 5.2.)

TABLE 5.2
1956 CONGRESSIONAL PREFERENCE

Party	Prefer Democrat (P)	Prefer Republicans (Q)	(P + Q)	(P − Q)
Democratic	.90	.07	.97	.83
Independent	.41	.38	.80	.03
Republican	.09	.85	.94	−.76
Average	.47	.43	.90	.04

The party trends are illustrated clearly in Figure 5.1, which gives the breakdown in Congressional voting preference by party for each of the five Congressional elections covered by the data bank.

A secondary factor in party preferences partially obscured by the over-all trends is worth noting. Party solidarity is higher in off-year elections than it is in Presidential years.

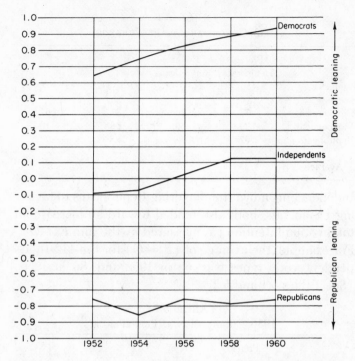

FIGURE 5.1 Party preferences in Congressional elections. Dependent variable in figure is $(P - Q)$, the difference between the percentage favoring Democrats for Congress and the percentage favoring Republicans. Thus increasing $(P - Q)$ indicates stronger Democratic leaning, while decreasing $(P - Q)$ indicates stronger Republican leanings.

Among the Republicans, where there was no over-all trend in $(P - Q)$, the greater off-year party polarization is clearly evident. Table 5.3 shows that polarization

TABLE 5.3
PARTY PREFERENCE IN CONGRESSIONAL ELECTIONS $(P - Q)$

Party	1952	1954	1956	1958	1960
Democrats	.67	.76	.83	.89	.92
Republicans	− .74	− .83	− .76	− .79	− .75

is greater in 1954, —.83, than in 1952 or 1956 and is also greater in 1958, —.79, than in 1956 or 1960. Among the Democrats, the greater off-year polarization is less evident because of the general increase in polarization over the entire period. Between successive Presidential elections, however, there is a greater increase in polarization going into the off-year, than in going from the off-year to the Presidential election. Polarization increased by .09 between 1951–1952 and 1953–1954 and by .07 from 1953–1954 to 1955–1956; in the next period, polarization increased by .06 from 1955–1956 to 1957–1958 and by only .03 from 1957–1958 to 1959.

Boundary effects as polarization approached 1.0 necessarily made the successive increases in polarization smaller. When this is taken into account, we see that it is clearly the off-years, 1954 and 1958, which show sharp increases in party polarization.

If the survey data analyzed here extended farther back in time, it would help confirm whether the party trends in Congressional voting shown in the 1950's were simply a return to normal voting behavior or the evolution of a new pattern. It is most likely that 1952 was the deviation and the subsequent years constituted a return to normal. Eisenhower was the first Republican President elected since 1928. Democrats and Independents, not Republicans, deviated from their normal patterns to elect him in 1952. It was Democrats and Independents who therefore showed a trend toward restored confidence in their usual judgment and restored party-line voting in subsequent years. In 1952, Eisenhower, a military hero who was equated in the public mind with peace, was the man of the hour. Considered almost bipartisan by many (after strong support for the nomination at the Democratic convention in 1948) he was "above" the "corruption" that was supposed to be shaking Washington at the time. The three issues of 1952—Communism, Corruption, and Korea—all favored Eisenhower and the Republicans. Comparatively good economic conditions allowed New Deal supporters to vote Republican without worrying about the threat of another 1929. After 1952 the Republicans gradually lost their charm, not for Republicans but for Democrats and Independents whose votes the Republicans needed in order to win.

SOUTHERN REPUBLICANISM AND EISENHOWER

Analysis of the Congressional voting data also indicates that Republican voting in Congressional elections may be ready to rise in the South. Upper-income Southerners in rural and town areas have shown marked drops in their support of the Democratic party. Outside the South, traditionally, Republican party strength has been related directly to income and to size of community; the higher the income and the smaller the community, the greater the Republican sentiment. In the South the pattern is partly reversed. It was in metropolitan areas in the South where, in the 1952 and 1956 Presidential elections, Republican strength appeared strongest. The small town was the stronghold of the Democratic tradition. It was only in the comparative freedom of the metropolis that Southerners expressed overt Republican preferences. Over the past eight years, however, the total expressed party preference in Congressional elections, $(P + Q)$, has decreased for only 2 groups of the 28 formed by a 2-way SES $\times$ region breakdown of voter-types: A + B rural Southerners and A + B town Southerners. These two groups outside the South would normally be mainly

Republican, but in the South they have not been able to break with their Democratic heritage. They have begun to do so in the form of declining to express any preference. These upper-class provincial Southerners have been under severe cross pressure of an avoidance-avoidance type and have responded with less-decided voting intentions. While the Republican vote preference of these groups remained virtually unchanged, the Democratic preference has decreased markedly.

The fact that Democratic solidarity is on the wane among well-to-do rural Southerners raises the probability of Republican Presidential or Congressional candidates winning in the South, even though these groups have shown no increased interest as yet in the Republican Congressional party. These groups are under cross pressures between their regional interests and their interests pertaining to their socioeconomic status. The cross pressure has resulted so far only in withdrawal from the Democratic Party. Ultimately, this situation will probably be resolved in favor of economic considerations, especially as the impact of industrialization on the South becomes more and more penetrating. In the meantime, however, these groups may vote less heavily, which will also aid Republicans by increasing the voting strength of the urban South where Republican strength is already growing.

Further evidence of the growth of a two-party system in the South is to be found in the questions dealing with support for President Eisenhower. They indicate that the development of latent Republican strength in the South was not due only to the appeal of Eisenhower, the man, but due also to other factors of a more stable nature.

There was a general Southern trend of reduced support for Eisenhower from 1952 to 1960. But differentially, those Southern socioeconomic-status groups that in other regions would normally be Republican showed smaller defection from Eisenhower over his terms in the White House than those groups that elsewhere would be Democratic. The A-level Metropolitans even show an increase in approval of Eisenhower, while normally Democratic lower-class groups turned away from him. This analysis rests upon differential trends. There was an over-all Southern disenchantment with Eisenhower, perhaps due to civil rights. But the Southern effect was not uniform across income levels and community sizes. The general Southern trend of disapproval of Eisenhower seemingly has superimposed upon it another trend by socioeconomic-status groupings which suggests the slow evolution of a Democrat-Republican strength pattern similar to that in other regions.

Table 5.4, showing the change in Southern support for Eisenhower for the seven socioeconomic-status groupings, reveals the evolution of a pattern of Republican strength by 1959 much more nationally typical than the pattern of 1954. The average Southern voter-type was more strongly pro-Eisenhower in 1953–1954, when $(P - Q)$ averaged $-.57$, than in 1959, when $(P - Q)$ averaged $-.48$. But within that trend there is a pattern for each level of community size whereby upper-income groups come to be the most solid supporters of Eisenhower. The table shows also that the growth of latent Republicanism suggested in the past section cannot be explained by increased attraction to Eisenhower. Neither among upper-class rural or upper-class town Southerners did support for Eisenhower increase. Yet for these two groups support for the Democratic party, as evidenced in Congressional preference, decreased. Latent Republicanism is there, but in the form still of a growing alienation from the Democratic party.

The magnitude of the differential changes in attitude toward Eisenhower for various Southern groups can be understood partly as a return to group norms. The

TABLE 5.4

AVERAGE BIENNIAL CHANGE IN CRITICISM OF EISENHOWER
IN SOUTH, BY SOCIOECONOMIC STATUS $(P - Q)$*

Area	Average Biennial Trend	1954 Level	1959 Level
A + B Rural	.002	−.53	−.52
C Rural	.006	−.43	−.41
A + B Town	.03	−.54	−.45
C Town	.05	−.56	−.41
A Metropolitan	−.01	−.59	−.62
B Metropolitan	.04	−.68	−.56
C Metropolitan	.08	−.64	−.40
Southern Average	.03	−.57	−.48

Note: * P indicates disapproval, Q approval (negative numbers are favorable to Eisenhower, positive numbers unfavorable).

A means professional, executive, and Managerial occupations; B means white-collar occupations, C means blue-collar occupations.

Urban refers to metropolitan areas exceeding 100,000 population; town refers to communities between 5,000 and 100,000; rural refers to places under 5,000.

more extremely a group deviates from its normal position in a period of unusual behavior, the more drastic will be its change as it returns to its normal view. This would explain why disapproval of Eisenhower increased most in the C Metropolitan and C Town levels, two groups that under the impact of the Eisenhower wave of 1952 swung far in deviation from their normal Southern Democratic proclivities. The A- and B-level voter-types whose support of Eisenhower was less abnormal changed less.

EISENHOWER SUPPORT AND COMMUNITY SIZE

The trend in support for Eisenhower varied markedly between rural, town, and metropolitan environments. Support for Eisenhower increased in the rural areas, and decreased in the large cities.

As Figure 5.2 shows, the strongest early approval for

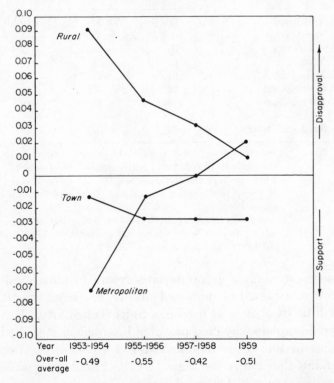

FIGURE 5.2 Differential support for Eisenhower Administration by community size.

Eisenhower as President was in metropolitan areas, and the weakest was in rural areas, in direct opposition to the national Republican pattern in which Republican strength is greatest in the rural and town areas. The changes in differential support show the steady return to normal in Eisenhower's support as the groups that most deviated from their normal position in 1953–1954 when Eisenhower first took office changed over time. In 1953–1954, for example, the average metropolitan voter-type was more strongly in support of Eisenhower than the average rural voter-type by .16. By 1959–1960, however, Eisenhower's support was higher among rural groups than among metropolitan ones.[2]

PARTY CLEAVAGE AND STEVENSON

One further trend illustrates the re-formation of party lines after their shattering by Eisenhower in

[2] Breaking down this same trend by socioeconomic status and religion we find different patterns for Catholics and Protestants. In the metropolitan areas these differences are not notable. Opposition to Eisenhower's policies increased among Protestants and Catholics alike and to a comparable degree. In the towns, however, where over-all support of Eisenhower declined very slightly, the decline was relatively substantial among Protestants while among Catholics support for Eisenhower increased. It is important to emphasize that these are differential, not absolute, trends. What this means is that in the towns the political difference between Catholics and Protestants was declining. Catholics who in the past had been predominantly Democratic were increasingly giving Republican replies on this issue-cluster. Political assimilation was taking place to a greater extent in the towns than in the metropolitan areas. In the rural areas the pattern was like that in the towns. Here again Catholics became more favorable to Eisenhower to a very substantial degree. Relatively speaking (but only relatively speaking) Protestants were becoming less favorable to him. The over-all trend in the rural areas (which are, of course, overwhelmingly Protestant) was strongly pro-Eisenhower but with the traditional religious difference tending to disappear over the eight years.

1952. That trend was in support for Stevenson. It is based on questions asking the respondents about their liking for Stevenson, not their vote intention in his two Presidential contests. The Stevenson issue-cluster was composed of questions from 28 surveys in the years 1953–1954, 1955–1956, 1957–1958, and 1959–1960. The variety of questions in the cluster was greater than for other issues and is a good example of a fabricated cluster composed of questions whose high intercorrelation indicates similar latent meanings. The questions are of two main types and deal with preferring Stevenson for President (but not in relation to the actual

TABLE 5.5
EISENHOWER AND STEVENSON ISSUE-CLUSTERS;
DIFFERENTIAL BIENNIAL TREND BY SOCIOECONOMIC
STATUS AND RELIGION*

	Protestants		Catholics	
	Eisenhower	Stevenson	Eisenhower	Stevenson
A + B Rural	− .001	− .002	− .045	.019
C Rural	.016	.005	− .064	.085
A + B Town	.016	.001	− .031	.055
C Town	.022	.042	− .033	− .027
A Metropolitan	.039	− .020	− .036	− .058
B Metropolitan	.018	− .038	.042	− .035
C Metropolitan	.028	− .036	.028	.009

* Dependent variable is $(P − Q)$.

Note: Negative trends favorable to Eisenhower, unfavorable to Stevenson. Positive trends favorable to Stevenson, unfavorable to Eisenhower.

election contest with Eisenhower) or general "liking" for Stevenson. Over the two terms of the Eisenhower era the image of Stevenson became a more partisan one. (See Table 5.5). Democrats moved relatively toward greater support of Stevenson while Republicans moved away from him; the Independents moved, relatively, very slightly away.

As Table 5.6 shows, at the very end, however, this

TABLE 5.6

DIFFERENTIAL SUPPORT FOR STEVENSON $(P - Q)$

	1954	1956	1958	1960
Democrats	.28	.39	.59	.55
Republicans	−.29	−.35	−.56	−.50
Independents	.01	−.04	−.03	−.04
Over-all Average	.28	.26	−.11	−.05

Note: Actual level of $(P - Q)$ for a group is equal to over-all average plus differential average. Thus 1954 levels of $(P - Q)$ for Democrats, Republicans, and Independents were .56(+.28 + .28), −.01(+.28 − .29), and .29(+.28 + .01).

trend stopped. Party cleavage in support of Stevenson had increased greatly from 1953–1954 to 1959–1960. But as 1960 approached and other political figures assumed the center of the Democratic stage as potential Presidential candidates, the party cleavage with regard to Stevenson decreased.

1952 AND 1956 PRESIDENTIAL RESPONSES

Nearly every Gallup Poll asks respondents whom they voted for in the last Presidential election and for whom they intend to vote in the next election. So the data bank includes surveys from before and after the 1952 and 1956 Presidential elections and provides us with an opportunity to compare "intended votes" with the votes recalled after the election.

Thus the issue-clusters dealing with the 1952 and 1956 elections provide data for observing whether there is a postelection bandwagon effect as many political scientists have claimed. Do people forget or falsify how they voted when they voted for the loser? Actually, our data are not consistent with the widely believed hypothesis that after an election, people join the bandwagon claiming that they supported the winner.[3] Post-election reporting of past votes is somewhat inaccurate, but for a different reason. Analysis of the variable $(P - Q)$ indicates that the major "before and after" trend in the election clusters can be accounted for by a kind of "return to normal." If a voter-type has a predisposition toward one party, then the heightened interest of the election campaign mobilizes the members of it toward party regularity. A campaign thus increases the group's distance from a 50-50 split. After the election, as the saliency of politics decreases, some individuals in the group slip away from the influence of the group norm and that is reflected in their remembered vote. The remembered vote of the voter-type, therefore, comes to differ less from the national norm. Democratic voter-types remember their vote as less one-sidedly Democratic than it was and Republican voter-types remember their vote as less one-sidedly Republican. This is shown by party trends in $(P - Q)$ from before to after each election.

One might try to explain the before-after change in Table 5.7 as the result of a "forgetting" process, whereby everyone tends slowly to forget how they voted as the election in question falls further into the past. Simple forgetting would introduce randomness in reporting and therefore would make voter-types regress toward the national average. However, simple forgetting would continue to have an increasing effect over time. In fact, however, the reported vote 3 to 4

---

[3] Of course, a bandwagon effect operative over only a short time period following the election might escape detection in our data, which is spread thinly over a long time period.

years after the elections is the same as the reported vote 1 to 2 years after the elections.

The figures in Table 5.7 can best be explained with reference to the political campaign and the heightened saliency of party competition that it brings. The campaign brings politics to a higher level of public attention than normal, and causes many people who normally take little or no interest in politics and parties to assume the partisan passions of their own voter-type.

TABLE 5.7
1952 AND 1956 PRESIDENTIAL ELECTIONS TRENDS IN $(P - Q)$

Year	Democrats	Republicans	Independents
1952*	$-.16$	.12	.08
1956†	$-.07$	.13	.15

* Comparison of 1951–1952 with 1953–1954 replies.
† Comparison of 1955–1956 with 1957–1958 replies.
P = vote for Stevenson
Q = vote for Eisenhower

It is these people who account for most of the before-after changes in reports of their candidate preference. People with only slight political involvement who are predisposed toward one party are drawn further from the "neutral" state as election salience increases but are likely to misremember their position after the election is over and they have returned to political hibernation.[4]

This explanation implies that the greater the interest in politics between campaigns, the less change between intended and reported vote. The changes in $(P + Q)$ lend support to this theory. All socioeconomic-status groups show marked increases in persons who "don't

[4] Our remarks here are substantially consistent with the analysis by Philip Converse, "Information Flow and the Stability of Partisan Attitudes," *Public Opinion Quarterly*, Vol. 26 (1962), pp. 578–599.

know" whom they favored from before the campaign to afterward, but the difference in the amount of such change between professional and white-collar workers on the one hand and blue-collar workers on the other is marked. In both elections, as shown in Table 5.8, the

TABLE 5.8
CHANGE FROM BEFORE TO AFTER ELECTION IN
"DON'T KNOWS" BY SES LEVEL

	1952 Election	1956 Election
A + B	+.12	+.18
C	+.22	+.29

Note: "Don't knows" = $[1 - (P + Q)]$.

increase in "don't know" responses was highest among the blue-collar voter-types, and it is among these that political indifference is highest.

CONCLUSION

The trends that have been discussed here indicate relative stability in American political attitudes. The decade covered by our data bank witnessed the re-emergence in the American electorate of most of the traditional party orientations of the pre-Eisenhower era. In general, the trends discussed were steady over the time studied with no drastic year-to-year fluctuations. Only in the case of support for Stevenson, in fact, did the over-all index change markedly in absolute, as opposed to differential, terms. The growth of party solidarity in Congressional voting and the differential change in support for Eisenhower by community size in particular showed the main trend of public opinion in the fifties to be the gradual return of a more normal

162

pattern of public opinion after the major disruption of 1952. Only in the case of the South, where the possibility appeared of a build-up in latent Republicanism, did the trends reveal what could be the development of a different political picture than before 1951–1952.

6

A Postscript on the 1964 Election

This book first appeared half a year before the elections of 1964. Soon there was another election to report, another opportunity to test the methods and conclusions that we had offered. This chapter is an addendum reporting on a second trial carried out during the 1964 campaign.

No two Presidential elections are alike. To the historian every election is a novel rearrangement of persistent elements of national life. To the politician every election is another risky play in a continuing game. To the political scientist every election is a fresh experiment against which to test his theories. The 1964 election campaign, so different from that of 1960 — indeed so different from that of any other in American history — was a challenge. How would our methods work in a second election and what would they show about the dynamics of the vote?

As we write this chapter just two weeks after election day three conclusions stand out:

1. Our basic simulation procedure worked. Once more, with data none of which was newer than two years old, we were able to simulate the outcome on the basis of cross-pressure theory with a degree of accuracy substantially like that in 1960.

2. The dynamics of 1964 were utterly different from those of 1960. In contrast to 1960, when group identification (i.e., the religious issue) was the main variable in accounting for votes and vote switches, in 1964 it was attitudes on issues that largely accounted for Johnson's landslide triumph over Goldwater. To a degree rare in American politics, policy issues, not group loyalties, dominated the campaign. The three issues that determined votes in 1964 were civil rights, nuclear responsibility, and social welfare legislation. (These issues often reduced to negative allegations about Goldwater's personality so that it is impossible to rule out personality as also a major campaign factor.)

3. While the methods we have used have proved effective once more, we have also by these two experiences identified certain potential improvements. In any future operation we would now know how to advance the art of election simulation further, taking full advantage of the increased capacity of modern computers.

MODELS OF THE 1964 CAMPAIGN

The 1964 campaign was special in many ways. It was record breaking in the size of President Johnson's victory. Viewed from the future it may turn out to be one of those rare elections that realign the political structure of the country. Such elections are few and far between. With the election of Jefferson in 1800, the Federalist Party disappeared; with the election of 1860, the Whigs disappeared; in the election of 1936, the majority Republican Party became the minority party as it remains to this day. The election of 1964 could turn out to be another such landmark election.

It is premature to assume that after its bruising defeat the Republican Party, captured by an extreme wing, will vanish from the American political scene. More likely the G.O.P. will experience a painful purging and recovery. But whatever happens, it will never be the same party that it was before Barry Goldwater launched his crusade to prove that he would rather be right than President.

Seldom before has a national major party candidate put so high a value on upholding moral and ideological principles rather than representing his people. The standard strategy of candidates in the American two-party system is to seize the well-peopled middle ground. The candidate in some sense holds captive those followers who are extremists of his own wing of opinion; they have little choice but to vote for him even if he is more moderate than they are. To win, however, he must demonstrate to the moderates among the voters that he is squarely in the middle with them. The old European shibboleth that American politics offers the voter no choice since the parties are but Tweedledum and Tweedledee turns the situation on its head. The parties speak like Tweedledum and Tweedledee because each tries to represent that choice that the voters have already made. Far from denying expression to the voter, the two parties are typically so anxious to serve the views of the mass that the job of representing the voters has already been done well before the candidate presents himself as a spokesman for that middle view in the debate. But not Barry Goldwater. He offered the voters a choice of something it was clear they did not want. The result was a landslide in which millions of voters broke from their traditional party, many for the first time in their lives.

Such broad outlines of the 1964 campaign were visible to any perceptive journalist, politician, or political scientist well before election day. The job in our 1964 pre-election simulation was to represent these dynamics in sufficient detail and with sufficient rigor to make it possible to calculate how different portions of the American electorate would respond to the campaign on election day. We report in the pages to follow three closely related simulations which we shall designate a five-factor, a six-factor, and a seven-factor simulation, and which successively approximate more fully what happened on election day.

The five-factor simulation was the basic simulation of which the others are but modifications. The five-factor simulation postulated that voters would behave according to the following simple model:

All Democrats except those who were strongly opposed to civil rights would vote for Johnson.

All Republicans would vote for Goldwater except those who supported the Democratic side on at least two out of three issue-clusters, namely, civil rights, nuclear responsibility, and social welfare legislation.

All Independents who expressed a clear view on civil rights would vote accordingly; those without a clear view on civil rights would vote in accordance with their views on the other two clusters. In estimating how Independents might divide we would disregard those Independents who expressed no views or whose views were evenly balanced.

We call this simple model a five-factor model because in addition to the three issue-clusters (civil rights, nuclear responsibility, and social welfare legislation) the outcome is also determined by two other factors: party identification and turnout. We expressed this model mathematically in the following equations.

For Democratic voter-types:

$$V_j = P_{36}(1.0 - Q_1 P_2)$$

For Republican voter-types:

$$V_j = P_{36}[P_1 P_3 + P_1 P_4(1 - P_3) + P_3 P_4(1 - P_1)]$$

For Independent voter-types:

$$V_j = P_{36} \left[\frac{\begin{matrix} P_1 + (1 - P_1 - Q_1)P_3 P_4 \\ + (1 - P_1 - Q_1)(1 - P_3 - Q_3)P_4 \\ + (1 - P_1 - Q_1)P_3(1 - P_4 - Q_4) \end{matrix}}{\begin{matrix} P_1 + (1 - P_1 - Q_1)P_3 P_4 \\ + (1 - P_1 - Q_1)(1 - P_3 - Q_3)P_4 \\ + (1 - P_1 - Q_1)P_3(1 - P_4 - Q_4) + Q_1 \end{matrix}} \right.$$

$$\left. \frac{\begin{matrix} + (1 - P_1 - Q_1)Q_3 Q_4 \\ + (1 - P_1 - Q_1)(1 - P_3 - Q_3)Q_4 \\ + (1 - P_1 - Q_1)Q_3(1 - P_4 - Q_4) \end{matrix}}{} \right]$$

Where

P_{36} = per cent of the voter-type with established voting habits (the turnout factor).

P_1 = per cent of the voter-type for civil rights.

Q_1 = per cent of the voter-type against civil rights.

P_2 = per cent of the voter-type feeling strongly about civil rights.

P_3 = per cent of the voter-type for nuclear restraint.

Q_3 = per cent of the voter-type against nuclear restraint.

P_4 = per cent of the voter-type for social welfare legislation.

Q_4 = per cent of the voter-type against social welfare legislation.

V_j = per cent of the voter-type predicted to vote Democratic.

The apparent simplicity of the model hides a couple of points of some importance. First of all note that we are assuming statistical independence among the three issues that determine the votes of Republicans and Independents. Unlike what we did in the 1960 simulation we did not assume any individual-by-individual correlation within voter-types between attitudes on civil rights, nuclear responsibility, and social welfare. A quick and admittedly superficial review of the data indicated that we could assume these attitudes to be orthogonal,[1] and so we did.

A second point that does not appear from reading the equations is that the issue-clusters that we used to index the three key issues were not raw survey data accepted at face values. They were modified values of issue-clusters in our data bank. Since each raw issue-cluster was composed from a large number of different questions asked at different times, it could be an arbitrary matter what proportion of the public was listed in the data bank as being for and against each issue. For example, on civil rights, a question on whether the respondent approves the Supreme Court's school decision produces many more answers on the pro side than does a question on whether the respondent favors open occupancy in housing.

In our 1960 simulation we made no allowances for this arbitrariness. We used the P's and Q's from the various issue-clusters just as they came from the data bank. It will be recalled for instance that we decided to use the exact proportion of respondents who indicated that

[1] There would, of course, be correlations across voter-types. Negroes and Jews would favor both civil rights and social welfare measures. But within a voter-type there need be no correlation across individuals. Those few Negroes who oppose integration are not necessarily the same individuals as those few Negroes who oppose higher minimum wages, for example.

they would hesitate to vote for a Catholic for President as our measure of anti-Catholicism. We noted, however, that if we had had some basis for doubting the relevance of that figure we might have chosen to weight it by a factor, for example, to assume that .8 or 1.1 of those who gave that answer really felt that way. In 1960 we could afford to take the data bank percentage at face value because we were not trying to estimate true vote percentages but only rank orders of states. The absolute level of P or Q in an issue-cluster did not make a sharp difference. In 1964, our simulation objective was more ambitions. We wanted to hit the right absolute level. That would clearly depend on the levels at which division occurred on the three attitudinal issues. We therefore had to give weights to the raw issue-clusters in some instances.

The civil rights clusters — both the cluster on attitudes toward civil rights and the cluster on the saliency of civil rights — were used exactly as they came out of the data bank. On social welfare legislation and on nuclear responsibility the over-all mean of the distribution was at a level that called for modification. The major social welfare issues in the 1964 campaign, such as social security, had a more popular base of support on the pro side than the typical social welfare issues represented in our cluster. Thus the data bank was negatively biased with respect to the relevant campaign themes. To adjust for this bias, 10 per cent of those respondents in each voter-type who had given replies classified as against social welfare legislation were reassigned to the pro side for our 1964 simulation. For a somewhat different reason, one third of all those respondents who had given replies against nuclear restraint were reassigned to a more pacific posture by 1964. In the latter case we regarded the reassignment as a true large shift rather than a correction for bias in the cluster content. The nuclear test ban treaty had been signed in October 1963, just a year before the election. The most recent data that we used were from at least a year before that. Many of the questions used in the nuclear responsibility cluster asked about fallout or about the banning of nuclear testing. Many individuals who opposed signing a test ban treaty with the Russians when the treaty was still a starry-eyed dream could be expected to favor it once it had become an accomplished fact, ratified by the Senate and apparently being observed.

Each of these adjustments fixed the mean distribution of views on the issue to about the same 60–40 level of support for the Democratic side that already held in regard to party affiliation and civil rights. It thus weighted the clusters roughly evenly. These parameter settings to a large extent determined the overall outcome of the simulation. The fact that this five-factor simulation gave Johnson 62 per cent of the national vote — he got 61.4 per cent — is not in itself a significant confirmation of the simulation procedure. That is largely a matter of good luck and good judgment in estimating the parameters. As in our 1960 simulation the real test is how close the simulation came to the results state by state.[2]

The five-factor simulation yields a result that is only moderately good, that is, a correlation with the November results of .52, and a median error of .061. The extent of deviation was no surprise. When we ran the five-factor simulation before the election we recognized that it did not represent all the relevant variables in the

[2] All statements about real November results refer to AP figures as of November 13, 1964. They are almost complete but not official. Thanks are due AP for specially providing them to us.

campaign, but it did serve to represent the action of the most interesting issues in the campaign. For prediction, however, it was obvious that certain other factors had to be considered too.

The obvious deficiency of the five-factor simulation lay in certain regional idiosyncrasies. In 1960 we had been able to represent the North rather well but the South much less well. In 1960 we also had made some serious errors in states that were not well classified within the regions to which they were assigned. Oklahoma, for example, was not a typical southern state, but we grouped it in the South. In general we had had trouble with the mountain states for with our system of simulated states we were representing states like Montana and Wyoming with data from such dissimilar places as California, and were representing states like Nebraska and North Dakota with data from such dissimilar states as Ohio and Michigan. The states of the western plains and the Rocky Mountains with their small population tended to be dominated by the larger population centers of the other parts of their assigned region. We realized, after 1960, that we could have done better had we had a separate mountain region.

As we looked at the five-factor simulation results before the election it became apparent that the same distortions were operating again in 1964 more sharply. Attitudes on the three crucial issues of the campaign that prevailed in the East, Midwest, or on the West Coast were producing simulation results for a region which by universal political judgment was considered likely to be Goldwater country.

It would have been easy to introduce an arbitrary geographic correction factor as we did for the South in 1960 in our best-fit simulation whereby it will be recalled, we moved 10 per cent of the white vote in the direction of the Republicans and 10 per cent of the Negro vote in the direction of the Democrats. We preferred, however, to find some rational basis if we could for an estimate of such a mountain and western plains state correction. A partly rational basis was provided by the extent to which we had erred in 1960 in estimating the western plains vote. We found that then, too, we had made them more Democratic than they should have been. Using the 1960 error as our estimate of the likely error in 1964 and rounding because of the basically arbitrary character of this adjustment, we took 5 percentage points off the estimate of the western plains states. Examination of the relationship in 1960 between those states and the mountain states led to a 10 percentage point correction in the latter. All of this was done in advance of the election.

Thus, our six-factor simulation was exactly as before with a geographic correction for a region that had special characteristics but that we had not singled out as a region in our data bank.

The six-factor simulation correlated with the actual outcome .63, and the median error was 3.6 per cent.

The seventh factor was also a geographic correction. In the 1964 campaign Johnson's name did not appear on the Alabama ballot. Voters in that state had no opportunity to cast a vote for him regardless of their preferences. Clearly it was preferences that we were simulating in our five- and six-factor simulations, not forced votes. Goldwater was fated to get 100 per cent of the Alabama vote, but we simulated what he would have gotten in a free election. The Mississippi situation was less clear-cut. Johnson electors were on the ballot. However the state Democratic Party unitedly swung to Goldwater. No prominent Democratic politician in the state remained loyal. The bolt was organized and complete. Johnson got 13 per cent of the Mississippi vote. We had, however, no basis for estimating that 13 per cent within the model we were using, for that represents not the free operation of the three issues on the minds of the voters but rather those voters rebelling against the organization's decisions.[3] The seventh factor was therefore simply to set Alabama and Mississippi at zero votes for Johnson.

This seven-factor simulation correlates with the actual election outcome .90, and the median error was 3.4 per cent.

All of these factors were known before the election, and our basic simulation was run then. Before the election we distributed to friends and colleagues the six-factor simulation. While the results in Alabama and Mississippi were fully determined in advance of the election and not a surprise, we did not then bother to change their simulated results to zero per cent for Johnson for we were reporting the consequences of a model, not an *ad hoc* prediction.

As in 1960 so in 1964 there were some other deviant simulation results in the South, this time Louisiana, North Carolina, and Arkansas. Before the election we had not concerned ourselves in any detail with the specific relationships among the Southern states because we felt on the basis of the 1960 results, that these relations, based on local political factors, were bound to be quite erratic, as indeed they were.

Table 6.1 shows the relationship of the six- and seven-factor simulations to the actual results. Figure 6.1 shows the relationship graphically for the states outside the South. The interesting thing at this point is to

---

[3] With Negroes providing nearly half of the Johnson votes, the nationally loyal white Democrats who broke with the state party were remarkably few.

TABLE 6.1   1964 SIMULATION RESULTS
(7-FACTOR SIMULATIONS SHOWING PERCENTAGE FOR JOHNSON)

State	Per Cent Predicted in 7-Factor Simulation	Actual Per Cent	State	Per Cent Predicted in 7-Factor Simulation	Actual Per Cent
Rhode Island	69	81	Illinois	62	60
Massachusetts	68	76	Montana	59	59
Maine	60	69	New Mexico	62	59
New York	69	68	Nevada	58	58
Connecticut	67	68	North Dakota	56	58
West Virginia	58	68	Arkansas	46	57
Michigan	62	67	Wyoming	58	56
Maryland	62	66	Indiana	60	56
Vermont	60	66	North Carolina	46	56
New Jersey	67	66	Oklahoma	47	56
Pennsylvania	64	65	South Dakota	55	56
Missouri	61	65	Tennessee	61	56
Kentucky	59	64	Utah	55	55
Minnesota	61	64	Kansas	55	54
Oregon	67	64	Virginia	49	54
New Hampshire	62	64	Nebraska	55	53
Texas	53	63	Florida	51	51
Ohio	61	63	Idaho	57	51
Washington	66	62	Arizona	59	50 −
Wisconsin	62	62	Georgia	48	46
Iowa	60	62	Louisiana	57	44
Colorado	57	61	South Carolina	48	41
Delaware	64	61	Mississippi	0*	13
California	68	60	Alabama	0*	0

* In 6-factor simulation Mississippi 48, Alabama 49.

examine the deviations of the actual from the predicted vote. What was going on within the electorate in 1964 that is not already taken into account in our simulation model? What other factors were in operation besides the ones already considered?

There was clearly a native son factor. Arizona went for Goldwater more markedly than our issue-based simulation predicted, and Texas more strongly for Johnson.

Two of the other cases of large deviation are noteworthy. While California did give 60 per cent of its vote to President Johnson, it was much more Republican than the balance of party and issue preferences in that state would have led us to expect. Our interpretation is

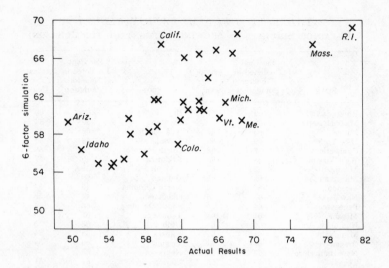

FIGURE 6.1 Comparison of the 6-factor simulation with the actual percentage vote for Johnson in the states outside the South.

that California was one state where a real backlash developed. In the same election that Johnson won by a wide margin, the voters of California by a two to one margin passed Proposition 14, which amended the state constitution to bar legislation for open-occupancy housing and simultaneously the California voters defeated liberal Pierre Salinger who opposed Proposition 14 by electing conservative Republican George Murphy who refused to oppose it. Johnson carried California despite a surge of anti-civil rights sentiment, but not by the margin he might have expected. With racial prejudice stirred up, many prejudiced voters resolved their cross pressures by voting for Goldwater.

Finally there was a reverse phenomenon in New England. New England rejected Goldwater in a way that went well beyond what could be explained by the factors in our model. We expected Rhode Island to be the most democratic state in the continent, and it was.[4] Massachusetts was second most Democratic in reality; we had expected it to be third. We gave Vermont and Maine to Johnson by large margins and that is how they voted. Nonetheless, as Figure 6.1 makes clear, New England went more strongly Democratic than the model accounts for.

Thus through our simulation and through the deviations from it, the dynamics of the election of 1964 become fairly clear. Some special forces were in operation in New England, in California, in Texas and Arizona, in Alabama and Mississippi, indeed such local variations were more important than they had been four years earlier. These variations modified a powerful surge of ideological voting against Goldwater, the candidate who first injected ideological issues into the campaign. He lost votes on civil rights, nuclear policy, and social issues. Johnson lost only segregationists.

Indeed the most dramatic aspect of the 1964 election was the extent to which it was genuinely dominated by issues rather than by social stratification. On the surface the reverse was the case. The press commented often that the speeches in this ostensibly ideological campaign failed to pose any issues clearly. There was little high-level discussion of issues. But whatever the contents of campaign output, the voters to an extraordinary extent did cast their votes on issues, or at least on personalistic encapsulations of the issues, such as Johnson's "restraint" versus Goldwater's "recklessness." This has not been true in all elections. When Lazarsfeld, Berelson, and Gaudet studied the 1940 election, they found that they could predict a man's vote with a high degree of confidence from a few of his social character-

[4] Recall that Hawaii and Alaska are not included in our data bank or results.

177

istics such as religion, rural-urban residence, and socio-economic status. Issue stands were not very predictive. In 1960 a man's vote could be predicted quite well on the basis of his party and religion. That was so only to a very limited extent in 1964. True, the Negro vote went overwhelmingly for Johnson, as did the Jewish vote. But these were exceptions. Johnson succeeded in creating a truly national coalition. He became, as he said he wanted to be, the candidate of all sorts of Americans. Probably never in recent years has there been so small a spread between the votes of business, professional people, white-collar people on the one hand and blue-collar people on the other. In our simulation only 7 percentage points separate the Democratic vote among blue-collar voters from that among the A and B SES people. Both gave their majority to Johnson. What determined a person's vote was not his social status as much as how he felt about the candidates' positions and styles on national issues. One consequence of the serious attention that the public paid to issues was an enormous number of defections from party ties. The opinion pollers can report to us from empirical surveys what number of Democrats voted for Goldwater and what number of Republicans voted for Johnson. Our simulation also provides an estimate. Altogether we find 17 per cent of Democrats voting for Goldwater, but these include many in the South. In the East and in the whole North we find 10 per cent of Democrats defecting to Goldwater. Among Republicans the defections were massive. More than a third of all Republicans voted for Johnson according to our simulation.

In 1960, it will be recalled, we found that of seventy million voters one in ten had switched from their expectable vote on the religious issue. In 1964 the number of voters who broke with their normal party vote on issue grounds was one in five or six.

Throughout this discussion we have stressed the vast differences between the election in 1960 and the election in 1964. The simplest evidence of that difference is the fact that the correlation state by state between these two elections actually comes out negative, −.16. That testifies how great the variety of outcomes a single stable attitude system of the American people can provide. Even though the attitudes are stable enough so that with two- to ten-year-old data we can accurately calculate how people will vote, two such different elections can yet emerge.

LESSONS FROM THE PAST AND FOR THE FUTURE

The 1964 simulation project reported in this chapter was conducted in a very modest fashion. It was not sponsored research. Unlike the unusual 1960 situation in which the Democratic Party launched the research project a year in advance, the Democratic Party had virtually no systematic research program in 1964. A number of state surveys were done but without much planning or coordination. Whatever was done was done in the very last months of the campaign.

The 1964 simulation was undertaken by the authors for scientific purposes. Its objective was to confirm the procedure we had developed four years earlier. Because it was unsponsored, the 1964 research had to be done on a minimum budget, but also it could be done in a university environment. We conducted the research at M.I.T. where available free computer time made the study possible.[5] To bring our data bank up to date, we obtained from the Roper Public Opinion Research Center twenty surveys conducted in the years 1960 through 1962. These additions to the data bank were particularly important because they brought the civil rights and nuclear responsibility issues into the current context.

For simplicity and economy we adopted the rule that we would not change our basic 1960 system in any way. We used existing computer programs from the 1960 study. We made no changes such as, for example, redefining regions or voter-types. All of these simplifying self-limitations were adopted in accordance with our narrow purpose of replicating the 1960 study.

By now, however, we know many ways in which to improve such election simulations. If we by any chance repeat such research in 1968, it will be done differently in many respects.

In 1960 we worked with an IBM 704 computer, large and fast in that day but inflexible and slow by present standards. We did not then feel it practical to repeatedly process 130,000 cards. So we obliged ourselves to group all responses into distributions within 480 voter-types. The development of modern computers has made this transformation of individuals into summary groups an unnecessary simplification. Today we could store and rapidly retrieve that many or more individual interviews. If we were programming the operation now we could keep the individual interviews available and regroup them at will into varying voter-

[5] Thanks are due to Project MAC and its time-sharing computer system for extraordinary facilities. Project MAC is an M.I.T. research project sponsored by the Advanced Research Projects Agency of the Department of Defense, under the Office of Naval Research, Contract No. Nonr-4102(01). Mr. Noel Morris conducted the computer operations.

types according to the character of the issues. We could change the definitions of regions at will (which we would have liked to do) or include age or ethnicity in the definition of the voter-types on some occasions. More accurately, we could include ethnicity to the extent that a national origin question was asked of the original survey respondent and the data stored on his card. The one thing that cannot be done in a secondary analysis such as our simulation is to provide raw information that the survey researcher did not originally ask. We are often deprived of information on such politically important matters as national origin because many polls do not ask it. Occasionally, however, the question is there. In a flexible simulation system which permitted the creation of voter-types at will, we could retain all such occasional information to be used whenever it was there and relevant.

By retaining the information from each original survey card in a fashion for rapid retrieval and reuse, we could also retain the freedom to modify and remake issue-clusters. By reference to the original information on individuals, we could estimate better any correlation that might exist between variables and regroup them when needed. Our assumption in the 1964 simulation that attitudes on civil rights, nuclear responsibility, and social welfare legislation were largely uncorrelated proved to be a satisfactory one. We would have been better off, however, if we could have tested the relationship among questions on these three topics wherever they co-occurred within our data bank. We could then with much more confidence have calculated the number of Republicans disagreeing with Goldwater on at least two issues.

In short, we are moving into an era where the limitation on data bank operations will no longer be computer capacity but only the imagination of the researcher and the extent to which raw data are available to him. For survey researchers this new age of rich data covering varieties of groups and long time series promises to bring a revolution in our ability to understand politics. For politicians and for citizens it promises the prospect of a more intelligent understanding of the civic process. The tools that we are now developing can handle models of a complexity adequate to the complexity of the political process itself. Not every election will be as intricate as that of 1964, but they will all be different and challenging to understand.

Glossary

ADDITIVITY Here is an example: *Democrats* are 45 points more favorable to Adlai Stevenson than are average voters; *urban* voters are 1 point more favorable to him than are average voters. If urban Democrats are 46 points (45 + 1) more favorable than average, then we have additivity of the two variables.

ANALYSIS OF VARIANCE A statistical technique for making estimates of how much effect each of several variables (separately and in combination) has on the variation within a set of measurements. For example, we observe variation in Democratic vote according to many variables: rural versus urban, high versus low SES, Catholic versus Protestant. We wish to know the effect on the vote of each variable alone and in combinations.

APPROACH-APPROACH CONFLICT A situation of choice between two alternatives, both desirable.

AVOIDANCE-AVOIDANCE CONFLICT A situation of choice between two alternatives, both undersirable.

BANDWAGON EFFECT An alleged behavior among voters whereby they wish to be on the winning side and thus vote for the candidate they expect to win rather than the one they otherwise prefer. There is little evidence that such an effect plays any significant part in American elections. The counteracting underdog effect is at least as large.

CHI-SQUARE A statistical test to determine whether the magnitude of deviation of an observed distribution of frequencies from a predicted distribution is so large as to render the prediction implausible.

CODE BOOK When a questionnaire is to be tabulated, each of the possible answers is represented on an IBM card by a certain punch (or punches) in a certain column (or columns). The code book specifies the particular punches and column numbers representing each possible answer. It is the document that is needed to interpret the punches on the IBM cards.

COMPUTER A computer is a device which performs arithmetic and logical operations on a body of stored data (or other symbols) in a sequence specified by a stored program. It differs from the familiar desk calculator in that at the beginning of a job one can store in it the series of numbers on which one is going to work and also the various operations that one wishes to have performed.

COMPUTER PROGRAM The instructions that tell the computer what sequences of operations to perform. Programs are written in special languages consisting of "commands" which trigger electronic responses appropriate to operations such as ADD, COMPARE, STORE, etc. The program is stored in the computer's memory at the start of a job.

CORRELATION A statistical measure of how closely two variables vary with each other. If all 6 foot men weighed 200 pounds, all 5′ 11″ men weighed 195 pounds, all 5′ 10″ men weighed 190 pounds and so on down 5 pounds with each inch, there would be a perfect correlation of height and weight.

There are many measures of correlation, one of the most used being the product moment coefficient for which the conventional symbol is r. A perfect correlation is ordinarily represented by $r = 1.0$. Actually there is not a perfect correlation of height and weight, but there is some correlation. Tall men do tend to weigh more than short ones. If there were no relationship, the coefficient of correlation would be $r = 0$. If there were a perfect inverse relationship, r would equal -1.0. So the range of correlation coefficients is from $+1.0$ to -1.0, plus meaning variable 1 goes up with variable 2, minus meaning variable 1 goes up as 2 goes down. The absolute size of the correlation coefficient tells us how nearly perfect the relationship, while the sign tells us if it is direct or inverse. In this book we have used some other measures besides *r* to get at the same notion of relationship.

CROSS PRESSURE   A situation in which a voter has some factors pushing him to vote one way and some factors pushing him to vote another way. In the original work in which the concept was formulated, Berelson, Lazarsfeld, and McPhee had in mind the special situation in which people close to the voter were inclined to the opposite party from his own natural one.

DATA BASE or DATA BANK   A set of facts or statistics available to be stored in the memory of a computer, which can then be used by the computer in calculations. The main data bank in this study was a set of tabulations of replies to public opinion polls.

DIFFERENTIAL TRENDS   Any change over time has to be measured against some standard. Most often the standard is an earlier measurement. Sometimes there are problems of comparability between earlier and later measurements. For example, if students take a midterm and final exam the questions are not the same. If only one student took the exam there would be no sure way of saying he did worse or better on the final than on the midterm. If a whole class takes the exams, however, we can say for any student whether he went up or down relative to the class average. That is a differential trend.

F-TEST   A test of the statistical significance of differences between two estimated variances in a sample of observations. The ratio of the two estimated variances is the F-ratio. Tables of the F-distribution give the maximum values of the F-ratio consistent with the hypothesis that the variances are equal in the entire population of observations.

GUTTMAN SCALE   A set of items is said to scale when they measure the same underlying variable but at different levels of intensity. In that situation if the answer to a less intense item is positive then so should be the answers to all the more intense ones. For example questions on giving Negroes the vote, or having them as neighbors, and on having them marry into one's family presumably form a scale. A person unwilling to concede the vote would be unlikely to accept a Negro into his family. Series of that type are said to scale.

INDEPENDENTS   Persons who belong to no political party. In this book classification of a person as an Independent means that he told the poll interviewer that he was neither a Republican nor a Democrat.

INTERACTION   When used in its statistical sense, this term indicates departure from additivity (q.v.). If, for example, young people are more likely to be Democrats than old ones, and if Southerners are more likely to be Democrats than are Northerners, but old Southerners are more likely to be Democrats than young Southerners, then there is an interaction. The effect of age depends upon (interacts with) region.

ISSUE-CLUSTER   In this study, to measure attitudes on any one issue we used a multiplicity of questions asked at different points in time. Thus a variety of specific questions about specific issues of foreign policy were all put together as dealing with the single issue-cluster of hard line versus soft line on relations with the Soviets. Some combinations of questions which we labeled issue-clusters actually did not deal with policy issues but rather with such matters as vote turnout. We constructed 52 issue-clusters in 1960.

LATENT ATTITUDE   The underlying attitude that may be tapped by a variety of questions. If questions scale (see Guttman scale) or if they are legitimately put together in a single issue-cluster, it is because they are all to some degree getting at the same latent attitude.

MAIN EFFECTS   In an analysis of variance (q.v.), those effects attributable to the variables taken one at a time, not together. Thus in this study main effects include sex, region, party, etc., and not their combinations (e.g., Southern Democrats).

MASTER MATRIX   The public opinion poll data in this study were tabulated by biennial election periods: 1951–1952, 1953–1954, 1955–1956, 1957–1958, 1959–1960, 1961–1962. These sets of data were then added together to produce grand totals for longer periods. There was a 4-period master matrix covering 1951–1958 and a 5-period master matrix covering 1951–1960. Those were used in the 1960 study. Then in the 1964 study the sixth period was added.

MATRIX   Any table in which the rows represent one variable, the columns another, and the cells contain some measure on the combination represented by the row-column intersection.

MEDIAN If a series of measures of a variable are placed in rank order, the measure in the middle of the ordering is the median. In the series (5672, 356, 3, 2, 1) the median is 3.

MEAN-SQUARED DEVIATION See VARIANCE.

ORDER OF INTERACTIONS In analysis of variance, after the main effects (q.v.) there are the first-order interactions which concern that part of the variance accounted for by the effects of two variables (e.g., Southern Democrats). Then there are second-order interactions which would concern joint effects of three variables, etc.

PARAMETER In simulation models a parameter is an intially unknown quantity which must be used in the calculations. In special cases, parameters may sometimes be estimated statistically from data appropriate to this purpose. In other cases, the parameters are given quantitative values by the programmer.

POLARIZATION In politics, polarization of attitudes is said to occur when most people hold attitudes at one or the other extreme of the distribution rather than in the middle.

PREDISPOSITIONAL FACTORS These are factors which make it likely that a person will accept a certain view, and which are operative before he has been exposed to that view. Thus Democrats are predisposed to think well of a Democratic candidate even before knowing anything about him.

ROOT-MEAN-SQUARE ERROR A measure of the deviation of a set of estimates from the true values. It is the square root of the variance (q.v.).

SALIENCE Social scientists generally apply this term to issues or attitudes. It refers to the degree of a person's concern and involvement in it. If an issue is important to a person and so in the forefront of his attention, it has high salience for him.

SAMPLE Out of a total population (the universe) a subset (the sample) is observed in order to estimate some value for the universe as a whole. To make a good estimate it is essential that the sample be in some sense representative of the universe. The sample can be made into a cross section of the whole by quota sampling, i.e., by requiring inclusion of major types of people in the sample in proportion to their occurrence in the universe. The sample can also be made representative by probability sampling, i.e., by drawing individuals in some random way such that each individual has an equal chance of being picked.

SAMPLE POINT   Few public opinion poll samples are true probability samples, for the cost of such is prohibitive. More often a series of geographic locations is randomly picked, but once a point is picked, a cluster of interviews is taken in the vicinity in some set fashion. Furthermore, survey organizations do not pick fresh sample points for each survey. They have to have interviewers residing in or near each sample point. Thus the pollers tend to keep their first broad selection of sample points constant for long periods.

SENSITIVITY TESTING   The output of a computer simulation depends not only on the model used and the data fed in but also on the settings given the parameters (q.v.). By running the simulation a number of times with different parameter settings one can test how sensitive the conclusions are to the parameter estimates.

SECONDARY ANALYSIS   When data collected for one purpose are later reanalyzed for another purpose, that is called secondary analysis. The present study is a secondary analysis of public opinion poll data.

SES   See SOCIOECONOMIC STATUS.

SIMULATION   Defined broadly, simulation is any attempt to model a system in such fashion that the changes the main system goes through are imitated by the behavior of the model. Defined that broadly simulation includes "war gaming," building of engineering models, and many other things. Here, however, we are describing a computer simulation. A computer simulation is a programming and running of a computer such as to make symbols in the computer's memory change in ways that presumably correspond to the changes in the system being simulated.

SIMULMATICS   A name coined for the project here described, the Simulmatics Project, and applied to the corporation which conducted it, the Simulmatics Corporation. The name was chosen for its suggestion of various appropriate words including simulation.

SOCIOECONOMIC STATUS Income, education, occupation, caste, ethnicity, respect, and various other factors all tend to distinguish high-status from low-status people in any society. These various indices are usually highly correlated with each other. Thus different researchers using different principles of classification (e.g., income or occupation) will in effect sort people out in much the same way. Lumping various indices together, one can call the ranking socioeconomic status. In the present study the basic breakdown was by occupation — business and professional versus white collar versus blue collar — but other indices of SES could be used if available.

STOCHASTIC PROCESS A process is stochastic if it is subject to random forces over a set of time points and one can only specify its outcomes in probabilistic terms. Actually, of course, it is often the model of reality that is stochastic rather than reality itself. For example, observation of a long series of digits dialed on a telephone by a stranger might permit one to construct a stochastic model of the occurrence of particular digits. To the person dialing, however, the pattern of digits is well determined, not random.

SUMMATE See ADDITIVITY.

SYNTHETIC STATE In the present study we analyzed voting behavior state by state. We did not, however, use data from within each state directly. Instead, we created "synthetic states" by calculating the proportion of voters in each state who belonged to each voting type and then estimating what the political behavior of the state should be from the types of voters who made it up.

TIME SERIES An array which states the value of some variable at a series of successive points in time. Trends are identified through time-series data. Each of our issue-clusters that spread over more than one biennial matrix provided a time series.

TURNOUT Turnout refers to the proportion of people who vote. The outcome of many elections is determined by which people get to the polls as much as or more than by conversions of any voters.

VARIANCE The average squared difference between a set of observations and a set of prespecified or predicted values for those observations. In the simplest interpretation, all prespecified values are equal to a single mean value, and the variance is then the average squared deviation of a set of observations around the mean value. In the analysis of variance (q.v.), the interpretation of variance differs, depending upon whether one is speaking of main effects (q.v.) or interactions (q.v.).

VOTER-TYPE The 160,000 interview respondents whose replies we analyzed were grouped into 480 voter-types. Each type was defined by SES, sex, religion, ethnicity, city size, region, and party.

Index